Multiple Scattering Theory

Electronic structure of solids

Multiple Scattering Theory

Electronic structure of solids

J S Faulkner
Florida Atlantic University, Boca Raton, FL, USA

G Malcolm Stocks
Oak Ridge National Laboratory, Oak Ridge, TN, USA

Yang Wang
Pittsburgh Supercomputing Center, Carnegie Mellon University, Pittsburgh, PA, USA

IOP Publishing, Bristol, UK

ISBN 978-0-7503-1490-9 (ebook)
ISBN 978-0-7503-1488-6 (print)
ISBN 978-0-7503-1489-3 (mobi)

DOI 10.1088/2053-2563/aae7d8

Version: 20181201

IOP Expanding Physics
ISSN 2053-2563 (online)
ISSN 2054-7315 (print)

British Library Cataloguing-in-Publication Data: A catalogue record for this book is available from the British Library.

Published by IOP Publishing, wholly owned by The Institute of Physics, London

IOP Publishing, Temple Circus, Temple Way, Bristol, BS1 6HG, UK

US Office: IOP Publishing, Inc., 190 North Independence Mall West, Suite 601, Philadelphia, PA 19106, USA

Contents

Author biographies

J S Faulkner, G Malcolm Stocks and Yang Wang

The authors' academic family tree
Jan Korringa (31 March 1915–9 October 2015) was a Dutch-American physicist best known for his discovery that the multiple-scattering theory could be used to calculate the electronic structure of solids. He was writing notes to the authors of this book in his famous barely legible script just weeks before his death at the age of one hundred, correcting their interpretation of his theory. From 1946 to 1952, Korringa was the protegé of Hendrik Kramers at the University of Leiden. Kramers was, in turn, the first protegé of Niels Bohr, so the connection with quantum mechanics started at the source.

Sam Faulkner was a PhD student of Korringa at the Ohio State University and wrote a thesis on the application of multiple scattering theory to disordered alloys. He became the leader of the Theory Group at the Oak Ridge National Laboratory. Malcolm Stocks came to ORNL as a post-doc in 1969 and collaborated with Faulkner on the first successful application of the coherent potential approximation to explain the electronic structure of a realistic alloy system. He spent the majority of his career at ORNL, applying multiple-scattering to explain phenomena in materials science and solid state physics. He recently retired as a Corporate Fellow and the leader of the Theory Group. Yang Wang was a student of Faulkner at the Florida Atlantic University and received a PhD in 1993. He then went to ORNL as a post-doc with Malcolm Stocks. During this period they developed, among other things, the locally self-consistent multiple scattering theory. Wang is a senior researcher at the Pittsburgh Supercomputing Center and directs PhD students at Carnegie Mellon University.

Over the years, the authors have continued to return periodically to ORNL and collaborate on research on various aspects of multiple scattering theory. Jan Korringa was a consultant at ORNL and interacted with the authors. Balazs Gyorffy, who directed numerous students at the University of Bristol, started his research in multiple scattering theory during his numerous visits to ORNL.

From the earliest days, the development and application of multiple scattering theory methods at ORNL has been supported by the Department of Energy and its predecessors, most recently the Office of Science, Basic Energy Sciences, for this the authors are extremely grateful.

IOP Publishing

Multiple Scattering Theory
Electronic structure of solids
J S Faulkner, G Malcolm Stocks and Yang Wang

Chapter 1

History of multiple scattering theory

The basic problem of condensed matter theory is to solve the Schrödinger equation for N bonding electrons. The resulting information about electronic states has made it possible to understand the properties of condensed matter and to predict the behavior of new materials. These new materials are the basis for the development of entirely new industries that have shaped the modern technology-based economy.

The justification for the present book is the conviction of the authors that, for certain applications, multiple scattering theory (MST) provides the best way to approach this problem. There are many reasons for this. If one applies MST correctly, it produces solutions of the Schrödinger equation that are numerically exact. Another advantage of MST is that the wave functions of the conduction or bonding electrons are automatically orthogonal to the bound state functions of the atoms. This seems like a technical point, but it is important to the accuracy of the results.

For ease of notation we write the Schrödinger equation in its nonrelativistic form, although it is standard to use the relativistic spin-dependent Dirac equation in actual calculations,

$$\left[-\sum_{i=1}^{N}\frac{\hbar^2}{2m_i}\nabla_i^2 + V(\mathbf{r}_1, \mathbf{r}_2, \ldots \mathbf{r}_N)\right]\Psi = E_{\text{total}}\Psi. \tag{1.1}$$

This partial differential equation in $3N$ variables, (1.1), is impossible to solve, but it can be simplified by making the Hartree–Fock approximation that leads to a set of one-electron equations. The Hartree–Fock approximation with configuration interactions is the best way to solve the Schrödinger equation for atoms and relatively small molecules. When the number of electrons in the system of interest is large, the Hartree–Fock approach becomes difficult to apply. More importantly, it emphasizes the wrong features of the electronic states, and frequently leads to incorrect conclusions. For large aggregates of atoms it is preferable to use approximations based on density functional theory (DFT) [1]. The local density approximation to DFT [2] leads to a set of one-electron equations

$$\left[-\frac{\hbar^2}{2m}\nabla^2 + v_{\text{eff}}(\mathbf{r})\right]\psi(\mathbf{r}) = E\psi(\mathbf{r}), \tag{1.2}$$

in which the effective potential, $v_{\text{eff}}(\mathbf{r})$, depends only on the charge density of the electrons, $\rho(\mathbf{r})$. As explained in references [1] and [2], there are limitations on the interpretation of E as a 'one-electron energy.'

After the N-electron Schrödinger equation has been reduced to a set of one-electron equations like (1.2), finding a solution is still a formidable problem. The technique of choice has been the Rayleigh–Ritz variational method [3], in which parameters in trial functions are adjusted to minimize the total energy. This approach leads to such band theories as the augmented plane wave method, the tight-binding method, the pseudopotential method, and others [4]. It came as a great surprise to the workers in the field when Korringa suggested a completely different approach for solving the one-electron equations. This approach, known as MST [5], arises from a picture of electrons scattering from one atom to another. The belief in the Rayleigh–Ritz method was so profound that many felt that Korringa's theory could not be correct, but a derivation of MST on the basis of the Kohn variational principle [6] reassured doubters that it is legitimate. It is now accepted that MST is an excellent method for solving one-electron Schrödinger equations. It has advantages over the techniques that use the Rayleigh–Ritz method, as will be seen in the following.

Korringa has written a brief history of MST [7], starting from a contribution made by Lord Rayleigh. In the process of explaining acoustical phenomena, Rayleigh invented techniques that have become essential tools for mathematical physicists, the variational method mentioned above being one example. The idea that led to MST appeared in a paper in which Lord Rayleigh derived the Lorentz–Lorenz relation for the dielectric constant for a two-dimensional periodic system [8]. Kasterin applied Rayleigh's idea to the scattering of acoustic waves by an array of spheres [9]. He constructed such an array in order to prove experimentally that his formulas are correct. He expressed the intention to apply his methods to the study of crystal optics, but there is no record of his doing so. In a series of famous papers, Ewald explained the interaction of electromagnetic waves with the atoms in a crystal using MST [10]. He derived seminal equations that are still used in MST, and laid the foundation for studying the structure of solids by x-ray diffraction.

In the theories of Rayleigh, Kasterin, and Ewald, there is an incoming wave on the collection of scatterers and an outgoing wave from it. On the occasion of his one hundredth birthday on March 31, 2015, Korringa wrote a note to the authors concerning the content of this chapter in which he gave some personal insights into the early stages of the development of MST. He described how it was on a train ride from the cities of Delft and Haarlem in the Netherlands that he had the epiphany that, when the number of scatterers approaches infinity, these equations could be used to obtain the stationary states of the system by searching for a solution when the incoming and outgoing waves are zero. The electrons scatter from one atom to another in a sequence that has neither a beginning nor an end. Korringa kept his result on his desk for a while in the hope that he could get someone to do calculations to illustrate his technique, but he finally gave up on that and published it in 1947. After he came to the United States, Korringa collaborated with Harrison [11] on the development of his ideas, but electronic computers had not reached a stage of development that made it possible to make much progress.

The timing of the paper by Kohn and Rostoker, deriving MST from a variational principle, was fortunate. During the 1950s, experimental techniques such as the de Haas–van Alphen effect were developed to the point that the Fermi surfaces of metals could be measured with great accuracy. There was a big push to develop new and improved methods for calculating the energy bands of solids. Free electron models gave a helpful first approximation to the measured Fermi surfaces of simple metals like sodium and aluminum [12]. It was in this environment that Ham and Segall [13] were inspired by the Kohn and Rostoker paper to develop procedures for calculating energy bands with MST using improved digital computers.

A group at Oak Ridge National Laboratory (ORNL) developed the MST method further, and wrote computer codes that fully automated the calculations [14]. The result is a band theory technique that became known as the Korringa–Kohn–Rostoker (KKR) method. They used their codes to calculate the electronic structure of solids, and made their computer codes available to other groups who wanted to apply the KKR method. The codes were later adapted to interpret the results of low energy electron diffraction (LEED) experiments [15]. A mathematical formalism based on MST was developed [16] and used to show the relation of the KKR with approximate band theory schemes such as the pivoted KKR, linearized augmented spherical wave (LASW) [17], and linearized muffin-tin orbital (LMTO) methods [18].

Band theory calculations demonstrated that the one-electron model using equation (1.2) gives excellent agreement with experiments. This came as a surprise to most of the leading theoretical physicists at the time, because they believed that the many-body effects that are not included in it would be much larger than the width of the energy bands.

Band theory calculations, by definition, apply only to solids with long-range atomic order. The question naturally arises, is it possible to calculate the electronic states for disordered solids? A method for doing this is called the coherent potential approximation (CPA) [19]. The first successful application of this theory was in the explanation of photoemission measurements on copper–nickel alloys [20]. The object of interest in the CPA is not the energy eigenvalues and wave functions that arise in band theory calculations but rather the one-electron Green's function. The only approach to the calculation of one-electron Green's functions that has been successful is based on MST [21]. The theory is called the KKR-CPA method, and it has been used to explain many properties of alloys, including magnetism and conductivity [22]. A useful formalism for one-electron Green's functions was developed and used to clarify some conceptual problems in the CPA theory [23].

The MST Green's function approach is the method of choice for treating another kind of disordered solid, impurities in an otherwise perfect crystal. Many such calculations have been carried out and used to explain the physics of such systems [24].

The CPA is useful for treating substitutional solid-solution alloys, but there are many forms of condensed matter without short-range order that do not fall in that category. A theoretical approach to the calculation of the one-electron Green's function within MST that can be applied to essentially any arrangement of atoms is called the locally self-consistent multiple scattering (LSMS) method [25]. The LSMS computer codes are so efficient that they are used for everything from systems with atomic and positional disorder to completely ordered crystals.

All of the KKR, CPA, and LSMS calculations originally made use of the muffin-tin approximation. In this approximation, the one-electron potential is assumed to be spherically symmetric within spheres that are centered on each nuclei. The spheres corresponding to different atoms do not overlap, and the potential is constant in the interstitial region between the spheres. The muffin-tin approximation is a good representation of the true potential for many metals, but it leads to errors in total energy calculations and calculations of the forces between atoms.

An effort was made to eliminate some of the shortcomings of the muffin-tin approximation by using spheres that enclose the same volume as the unit cell that contains an atom. The spheres overlap and it is not correct to apply MST to such a model. However, this atomic sphere approximation (ASA) is still in use because of its simplicity.

The present trend is to focus on full potential MST in which no shape approximation is made to the potential. This makes the calculations more difficult, but they are tractable because the speed of computers is increasing at the same time. Early efforts to do such calculations stop short of a true implementation of the MST approach [26, 27], but this has been achieved in a recent paper [28].

In the following chapters, the equations of multiple scattering theory will be derived. Modern computational techniques for using these equations in practical calculations of the properties of real materials are explained. A number of specific calculations will be described as examples of various aspects of the theory.

During the 1970s, it was argued that the computational methods obtained from the pure MST method described in this book require too much CPU time. For this reason, a number of 'linearized' band theory methods were suggested with acronyms such as LMTO, LAPW, etc. These methods have the disadvantage that they contain many hidden approximations at almost every stage, but this was considered to be a necessary affliction that must be borne in order to achieve the goal of practical calculations. Of course, CPU performance has improved at a rate that is similar to the prediction of Moore's law. Even more important is the introduction of massively parallel computer (MPP) architectures in which large numbers of CPUs work in unison, which are particularly well-suited for MST calculations. With these improvements, any introduction of an approximation must be justified by an argument that is not based on computational efficiency.

References

[1] Hohenberg P and Kohn W 1964 *Phys. Rev.* **136** B864

[2] Kohn W and Sham L J 1965 *Phys. Rev.* **140** A1133

[3] Rayleigh J W 1870 *Phil. Trans.* **161** 77
Ritz W 1908 *J.Reine Angew. Math.* **135** 1–61
[4] Slater J C 1965 *Quantum Theory of Molecules and Solids Volume 2: Symmetry and Energy Bands in Crystals* (New York: McGraw Hill)
[5] Korringa J 1947 *Physica* **13** 392
[6] Kohn W and Rostoker N 1954 *Phys. Rev.* **94** 1111
[7] Korringa J 1994 *Phys. Rep.* **238** 341
[8] Rayleigh L 1892 *Phil. Mag.* **34** 481
[9] Kasterin N P 1897 *Proc. Amsterdam* **6** 460
[10] Ewald P P 1916 *Ann. Phys., Lpz.* **49** 1
[11] Harrison R J 1951 *Phys. Rev.* **84** 377
[12] Harrison W A 1960 *Phys. Rev.* **118** 1190
[13] Ham F S and Segall B 1961 *Phys. Rev.* **124** 1786
[14] Faulkner J S, Davis H L and Joy H W 1967 *Phys. Rev.* **161** 656
[15] Noonan J R and Davis H L 1984 *Phys. Rev.* B **29** 4349
[16] Faulkner J S 1979 *Phys. Rev.* B **19** 6186
[17] Williams A R, Kubler J and Gelatt C D 1979 *Phys. Rev.* B **19** 6094
[18] Andersen O K 1975 *Phys. Rev.* B **12** 3060
[19] Soven P 1967 *Phys. Rev.* **156** 809
[20] Stocks G M, Williams R W and Faulkner J S 1971 *Phys. Rev. Lett.* **26** 253
[21] Stocks G M, Temmerman W M and Gyorffy B L 1978 *Phys. Rev. Lett.* **41** 339
[22] Faulkner J S 1982 *Progress in Materials Science* ed J W Christian, P Haasen and T B Massalski (Oxford: Pergamon) Vol. 27
[23] Faulkner J S and Stocks G M 1980 *Phys. Rev.* B **21** 3222
[24] Zeller R and Dederichs P H 1979 *Phys. Rev. Lett.* **42** 1713
[25] Wang Y, Stocks G M, Shelton W A, Nicholson D M, Szotek Z and Temmerman W M 1995 *Phys. Rev. Lett.* **75** 2867
[26] Dederichs P H, Drittler B and Zeller R 1992 *Application of Multiple Scattering Theory to Material Science* ed W H Butler, P H Dederichs, A Gonis and R L Weaver, *MRS Symp. Proc. No. 253* (Pittsburgh, PA: Material Research Society) p 185
[27] Zabloudil J, Hammerling R, Szunyogh L and Weinberger P 2005 *Electron Scattering in Solid Matter* (Berlin: Springer)
[28] Rusanu A, Stocks G M, Wang Y and Faulkner J S 2011 *Phys. Rev.* B **84** 035102

IOP Publishing

Multiple Scattering Theory

Electronic structure of solids

J S Faulkner, G Malcolm Stocks and Yang Wang

Chapter 2

Scattering theory

There are many excellent books on the subject of scattering theory. However, the topics in scattering theory that are needed in applications of MST are not covered very well in those texts. The primary interest in MST is the elastic scattering of electrons from potentials of finite range, $v(\mathbf{r})$. These are potentials which are zero outside of a finite volume, Ω. Many of the subtle mathematical arguments in the standard texts are unnecessary for such potentials, and the topics that are of interest are lost in the jumble of irrelevant material. In particular, modern developments in MST deal with potentials that not spherically symmetrical, so there must be an emphasis on developing practical methods for calculating the scattering matrices for such potentials using a digital computer.

2.1 Potential scattering

The scattering of a particle from a potential is different from other problems in quantum mechanics because the initial state, the free particle, has exactly the same energy as the final state, the particle traveling away from the scatterer. Logically and mathematically the only way to describe this process is for the incoming state to be a wave packet with a large enough distribution in momenta so that the particle can be reasonably well localized. The manipulation of wave packets is laborious, and the outcome of the analysis simply validates the solutions that are obtained from the Lippmann–Schwinger equations that are derived below.

The one-electron Schrödinger equation may be written in the Dirac notation

$$(E - H_0)\,|\psi\rangle = V\,|\psi\rangle, \tag{2.1}$$

where H_0 is the kinetic energy operator. The equation for the free-particle state vector is

$$(E - H_0)\,|\varphi\rangle = 0. \tag{2.2}$$

The solution of the homogeneous equation, equation (2.2), plus a particular solution of the inhomogeneous equation, equation (2.2), is

$$|\psi\rangle = |\varphi\rangle + (E - H_0)^{-1} V |\psi\rangle. \tag{2.3}$$

The algebra that leads to equation (2.3) is incorrect because E is an eigenvalue of H_0 and thus the inverse $(E - H_0)^{-1}$ doesn't exist. The mathematical trick that is used to get around this difficulty is to add a small imaginary part to the one-electron energy E. This leads to the equation

$$|\psi\rangle = |\varphi\rangle + G_0^+ V |\psi\rangle, \tag{2.4}$$

where

$$G_0^+ = \lim_{\varepsilon \to 0} (E - H_0 + i\varepsilon)^{-1}. \tag{2.5}$$

The imaginary part can approach zero from above or below the real axis. It will be shown that in order to satisfy the physical boundary condition that the second term represents an outgoing wave, one must approach zero through positive values.

Equation (2.4) is the Lippmann–Schwinger equation, and it is the starting point for modern treatments of scattering theory. Using the definition

$$\left(1 - G_0^+ V\right)^{-1} = 1 + G_0^+ V + G_0^+ V G_0^+ V + \ldots = 1 + G_0^+ T, \tag{2.6}$$

where

$$T = V + V G_0^+ V + V G_0^+ V G_0^+ V + \ldots = V\left(I + G_0^+ T\right), \tag{2.7}$$

it follows that another way to write the Lippmann–Schwinger equation is

$$|\psi\rangle = (1 + G_{0+} T) |\varphi\rangle. \tag{2.8}$$

2.2 Position representation

Putting the Lippmann–Schwinger equation in the position representation leads to

$$\langle \mathbf{r} | \psi \rangle = \langle \mathbf{r} | \varphi \rangle + \langle \mathbf{r} | G_{0+} T | \varphi \rangle, \tag{2.9}$$

or

$$\psi(\mathbf{r}) = \varphi(\mathbf{r}) + \iint G_0^+(E, \mathbf{r}, \mathbf{r}') T(\mathbf{r}', \mathbf{r}'') \varphi(\mathbf{r}'') d\mathbf{r}' d\mathbf{r}''. \tag{2.10}$$

In dimensionless units, the eigenvectors of the free-particle Hamiltonian H_0 are the solutions of the Helmholtz equation

$$[\nabla^2 + E]\varphi(\mathbf{r}) = 0. \tag{2.11}$$

A complete set of solutions of this equation are the functions

$$\varphi_{\mathbf{k}}(\mathbf{r}) = \langle \mathbf{r} | \mathbf{k} \rangle = \frac{1}{(2\pi)^{3/2}} e^{i\mathbf{k}\cdot\mathbf{r}}, \tag{2.12}$$

where we have written $E = k^2$. Using the resolution of the identity

$$\int |\mathbf{k}\rangle\langle\mathbf{k}|\, d\mathbf{k} = \mathbf{I}, \tag{2.13}$$

we obtain

$$G_0^+(E, \mathbf{r}, \mathbf{r}') = \iint \langle\mathbf{r}|\mathbf{k}\rangle\langle\mathbf{k}|G_0^+|\mathbf{k}'\rangle\langle\mathbf{k}'|\mathbf{r}'\rangle d\mathbf{k}d\mathbf{k}'. \tag{2.14}$$

This integral is evaluated using complex variable theory, and the contour in complex k-space must be closed in the upper half plane. When E is positive, the pole in the upper half plane is near $k = \sqrt{E}$, and

$$G_0^+(E, \mathbf{r}, \mathbf{r}') = \frac{1}{(2\pi)^3}\lim_{\varepsilon\to 0}\int \frac{e^{i\mathbf{k}\cdot(\mathbf{r}-\mathbf{r}')}}{(E - k^2 + i\varepsilon)}d\mathbf{k} = -\frac{\mathbf{1}}{4\pi}\frac{e^{i\alpha|\mathbf{r}-\mathbf{r}'|}}{|\mathbf{r} - \mathbf{r}'|}, \tag{2.15}$$

where $\alpha = \sqrt{E}$. Inserting this Green's function in the Lippmann–Schwinger equation gives an integral equation that is equivalent to the scattering equation

$$\psi_{\mathbf{k}}(\mathbf{r}) = \varphi_{\mathbf{k}}(\mathbf{r}) - \frac{1}{4\pi}\int \frac{e^{i\alpha|\mathbf{r}-\mathbf{r}'|}}{|\mathbf{r} - \mathbf{r}'|}V(\mathbf{r}')\psi_{\mathbf{k}}(\mathbf{r}')d\mathbf{r}'. \tag{2.16}$$

An alternative way to write the Lippmann–Schwinger equation is

$$\psi_{\mathbf{k}}(\mathbf{r}) = \varphi_{\mathbf{k}}(\mathbf{r}) - \frac{1}{4\pi}\iint \frac{e^{i\alpha|\mathbf{r}-\mathbf{r}'|}}{|\mathbf{r} - \mathbf{r}'|}T(\mathbf{r}', \mathbf{r}'')\varphi_{\mathbf{k}}(\mathbf{r}'')d\mathbf{r}'d\mathbf{r}''. \tag{2.17}$$

2.3 The classic scattering experiment

In a scattering experiment, the dimensions of the scatterer are typically small compared with those of the measuring apparatus. Mathematically, this means that $V(\mathbf{r}')$ is zero outside a small volume Ω. We are interested in the wave function at the position of the counter, so we will evaluate it for the case $|\mathbf{r}| \gg |\mathbf{r}'|$. In this limit, $|\mathbf{r} - \mathbf{r}'| \cong r - \frac{\mathbf{r}\cdot\mathbf{r}'}{r}$, and the Green's function is

$$G_0^+(E, \mathbf{r}, \mathbf{r}') \cong -\frac{\mathbf{1}}{4\pi}\frac{e^{i\alpha r}}{r}e^{-i\mathbf{k}'\cdot\mathbf{r}''}, \tag{2.18}$$

where

$$\mathbf{k}' = \frac{k\mathbf{r}}{r}, \tag{2.19}$$

is the k-vector pointing toward the outgoing particle. Thus,

$$\psi_{\mathbf{k}}(\mathbf{r}) = \frac{1}{(2\pi)^{3/2}}\left[e^{i\mathbf{k}\cdot\mathbf{r}} + \frac{e^{i\alpha r}}{r}f(\mathbf{k}, \mathbf{k}')\right], \tag{2.20}$$

where

$$f(\mathbf{k}, \mathbf{k}') = -2\pi^2\langle \mathbf{k}'| V | \psi_{\mathbf{k}}\rangle = -2\pi^2\langle \mathbf{k}'| T |\mathbf{k}\rangle. \tag{2.21}$$

The differential scattering cross section is defined by

$$\frac{d\sigma}{d\Omega}\Delta\Omega = \frac{\text{number of particles scattered into } \Delta\Omega \text{ per unit time}}{\text{number of incident particles per unit area and time}}. \tag{2.22}$$

Using the expression for the current $\mathbf{j} = \frac{\hbar}{m}\mathrm{Im}\psi^*\nabla\psi$, it follows that

$$\frac{d\sigma}{d\Omega} = |f(\mathbf{k}, \mathbf{k}')|^2. \tag{2.23}$$

The wave function can be made time dependent by multiplying it with an exponential

$$\psi_{\mathbf{k}}(\mathbf{r}, t) = \psi_{\mathbf{k}}(\mathbf{r})e^{-iEt}. \tag{2.24}$$

The time dependent version of equation (2.20) describes a free-particle wave approaching the scatterer, and an outgoing wave leaving it.

The scattering factor $f(\mathbf{k}, \mathbf{k}')$ is proportional to the t-matrix, as shown in equation (2.21). Since $|\mathbf{k}| = |\mathbf{k}'| = \sqrt{E}$, this is known as the on-the-energy-shell t-matrix.

2.4 Angular momentum expansion

The version of the Lippmann–Schwinger equation that is useful in MST is very different from equation (2.20). A spherical wave expansion can be used to describe the wave function as close to the scatterer as is necessary, and even inside the scatterer. The plane wave can be expanded in spherical waves

$$\varphi_{\mathbf{k}}(\mathbf{r}) = (2\pi)^{-3/2}e^{i\mathbf{k}\cdot\mathbf{r}} = \left(\frac{2}{\pi}\right)^{1/2}\sum_L i^l Y_L^*(\mathbf{k})Y_L(\mathbf{r})j_l(\alpha r), \tag{2.25}$$

and the free-particle Green's function can be written

$$G_0^+(E, \mathbf{r} - \mathbf{r}') = -\frac{1}{4\pi}\frac{e^{i\alpha|\mathbf{r}-\mathbf{r}'|}}{|\mathbf{r} - \mathbf{r}'|} = -i\alpha\sum_{L'} Y_{L'}(\mathbf{r})h_{l'}^+(\alpha r)j_{l'}(\alpha r')Y_{L'}^*(\mathbf{r}'), \tag{2.26}$$

so the scattering wave function in the equation can be written

$$\psi_{\mathbf{k}}(\mathbf{r}) = \left(\frac{2}{\pi}\right)^{1/2}\sum_L i^l Y_L^*(\mathbf{k})\psi_L(E, \mathbf{r}). \tag{2.27}$$

In this notation

$$\psi_L(E, \mathbf{r}) = Y_L(\mathbf{r})j_l(\alpha r) - i\alpha\sum_{L'} Y_{L'}(\mathbf{r})h_{l'}^+(\alpha r)t_{L', L}(E), \tag{2.28}$$

with

$$t_{L',L}(E) = \iint j_{l'}(\alpha r')Y^*_{L'}(\mathbf{r}')t(\mathbf{r}', \mathbf{r}'')Y_L(\mathbf{r}'')j_l(\alpha r'')d\mathbf{r}'d\mathbf{r}''. \tag{2.29}$$

The $\psi_L(E, \mathbf{r})$ are a complete set of solutions of the equation just like the $\psi_{\mathbf{k}}(E, \mathbf{r})$. It will be shown later that there are methods for calculating $t_{L',L}(E)$ for all potentials that will be used in MST.

Scattering theorists prefer to work with the s-matrix, which is related to the t-matrix by

$$\mathbf{S} = \mathbf{I} - i2\alpha\mathbf{T}. \tag{2.30}$$

Using this matrix, the preceding formula can be manipulated into the form

$$\psi_L(E, \mathbf{r}) = \frac{1}{2}\left[Y_L(\mathbf{r})h_l^-(\alpha r) + \sum_{L'} Y_{L'}(\mathbf{r})h_{l'}^+(\alpha r)S_{L'L}\right]. \tag{2.31}$$

2.5 Non-spherical potentials with finite domains

The potentials of interest in modern applications of MST are not spherical, but they are local and have a finite domain, Ω. Another way of putting this is that there is a finite radius A such that $v(\mathbf{r}) = 0$ for $|\mathbf{r}| > A$. This kind of potential is not treated in standard texts on scattering theory.

In order to extend the arguments of Jost to find the scattering matrices for non-spherical potentials, it is first necessary to obtain a solution to the Schrödinger equation that is regular at the origin. It can be shown by direct substitution that such a solution of the equation

$$[-\nabla^2 + v(\mathbf{r}) - E]\phi(E, \mathbf{r}) = 0, \tag{2.32}$$

is

$$\phi_{L_0}(E, \mathbf{r}) = Y_{L_0}(\hat{\mathbf{r}})j_{l_0}(\alpha r) + \int_{|\mathbf{r}'|\leqslant|\mathbf{r}|} K(E, \mathbf{r}, \mathbf{r}')v(\mathbf{r}')\phi_{L_0}(E, \mathbf{r}')d\mathbf{r}', \tag{2.33}$$

where the integration is over all $|\mathbf{r}'| \leqslant |\mathbf{r}|$. The kernel of this Volterra type integral equation is

$$K(E, \mathbf{r}, \mathbf{r}') = -\alpha\sum_L Y_L(\hat{\mathbf{r}})\left[j_l(\alpha r)n_l(\alpha r') - n_l(\alpha r)j_l(\alpha r')\right]Y^*{}_L(\hat{\mathbf{r}}'). \tag{2.34}$$

The function $\phi_{L_0}(E, \mathbf{r})$ is regular at the origin, and, in fact,

$$\lim_{|\mathbf{r}|\to 0} \phi_{L_0}(E, \mathbf{r}) \to Y_{L_0}(\hat{\mathbf{r}})j_{l_0}(\alpha r). \tag{2.35}$$

The subscript L_0 refers only to the angular momentum function associated with the initial condition. The function can indeed be expanded in spherical harmonics

$$\phi_{L_0}(E, \mathbf{r}) = \sum_L Y_L(\hat{\mathbf{r}})\phi_{LL_0}(E, r). \tag{2.36}$$

The function is also defined within the range of the potential as well as outside of it.

It can similarly be shown that

$$f_{L_1}^{\pm}(E, \mathbf{r}) = Y_{L_1}(\hat{\mathbf{r}})h_{l_1}^{\pm}(\alpha r) - \int_{|\mathbf{r}'|>|\mathbf{r}|} K(E, \mathbf{r}, \mathbf{r}')v(\mathbf{r}')f_{L_1}^{\pm}(E, \mathbf{r}')d\mathbf{r}', \tag{2.37}$$

where the integration is over all $|\mathbf{r}'| \geqslant |\mathbf{r}|$, are also solutions of the differential equation (2.32). When $|\mathbf{r}| > R_{bs}$, the bounding sphere radius for which $v(\mathbf{r}) = 0$,

$$f_{L_1}^{\pm}(E, \mathbf{r}) = Y_{L_1}(\hat{\mathbf{r}})h_{l_1}^{\pm}(\alpha r). \tag{2.38}$$

It is obvious that $f_{L_1}^{+}(E, \mathbf{r})$ and $f_{L_1}^{-}(E, \mathbf{r})$ are linearly independent, so it is possible to write

$$\phi_{L_0}(E, \mathbf{r}) = \frac{1}{2}\sum_{L_1}\left[f_{L_1}^{+}(E, \mathbf{r})a_{L_1L_0}(E) + f_{L_1}^{-}(E, \mathbf{r})b_{L_1L_0}(E)\right], \tag{2.39}$$

where $a_{L_1L_0}(E)$ and $b_{L_1L_0}(E)$ are independent of $\mathbf{r}$. We emphasize again that, by construction, the functions $\phi_{L_0}(E, \mathbf{r})$, $f_{L_1}^{+}(E, \mathbf{r})$, and $f_{L_1}^{-}(E, \mathbf{r})$ are all solutions of the differential equation for all $\mathbf{r}$. Another wave function that is a solution for all $\mathbf{r}$ is,

$$\psi_L(E, \mathbf{r}) = \frac{1}{2}\left\{f_L^{-}(E, \mathbf{r}) + \sum_{L_1, L_0}\left[f_{L_1}^{+}(E, \mathbf{r})a_{L_1L_0}(E)b_{L_0L}^{-1}(E)\right]\right\}. \tag{2.40}$$

Comparing this function in its asymptotic form for $r > R_{bs}$ with equation (2.31), it can be seen that the s-matrix is related to the $\mathbf{a}$ and $\mathbf{b}$ matrices,

$$\mathbf{S}(E) = \mathbf{ab}^{-1}. \tag{2.41}$$

The matrices $\mathbf{a}$ and $\mathbf{b}$ can be calculated from the integrals

$$a_{L_1L_0} = \delta_{L_1L_0} - i\alpha\lim_{\delta\to 0}\int_{\mathbf{r}\subset V_\delta^A} \mathbf{h}_{L_1}^{-}(E, \mathbf{r})V(\mathbf{r})\phi_{L_0}(E, \mathbf{r})d\mathbf{r}, \tag{2.42}$$

and,

$$b_{L_1L_0} = \delta_{L_1L_0} + i\alpha\lim_{\delta\to 0}\int_{\mathbf{r}\subset V_\delta^A} \mathbf{h}_{L_1}^{+}(E, \mathbf{r})V(\mathbf{r})\phi_{L_0}(E, \mathbf{r})d\mathbf{r}, \tag{2.43}$$

where $\mathbf{h}_{L_1}^{\pm}(E, \mathbf{r}) = h_{l_1}^{\pm}(\alpha r)Y_{L_1}^{*}(\hat{\mathbf{r}})$. Alternative expressions are

$$a_{L_1L_0} = -i\alpha\iiint_{S_A}\left\{\mathbf{h}_{L_1}^{-}(E, \mathbf{r}), \phi_{L_0}(E, \mathbf{r})\right\}.\, \mathbf{n}dS, \tag{2.44}$$

and,

$$b_{L_1L_0} = i\alpha\iiint_{S_A}\left\{\mathbf{h}_{L_1}^{+}(E, \mathbf{r}), \phi_{L_0}(E, \mathbf{r})\right\}.\, \mathbf{n}dS, \tag{2.45}$$

where $\left\{\mathbf{h}^{-}_{L_1}(E, \mathbf{r}), \phi_{L_0}(E, \mathbf{r})\right\}$ is the vector

$$\left\{\mathbf{h}^{-}_{L_1}(E, \mathbf{r}), \phi_{L_0}(E, \mathbf{r})\right\} = \mathbf{h}^{-}_{L_1}(E, \mathbf{r})\nabla\phi_{L_0}(E, \mathbf{r}) - \phi_{L_0}(E, \mathbf{r})\nabla\mathbf{h}^{-}_{L_1}(E, \mathbf{r}). \tag{2.46}$$

For reasons that will become clearer later on, we choose to define sine and cosine matrices **s** and **c** by

$$\begin{aligned} \mathbf{a} &= \mathbf{c} + i\mathbf{s} \\ \mathbf{b} &= \mathbf{c} - i\mathbf{s} \end{aligned}. \tag{2.47}$$

From the above

$$\begin{aligned} c_{L_1L_0} &= \delta_{L_1L_0} - \alpha\lim_{\delta\to 0}\int_{\mathbf{r}\subset V_\delta^A} \mathbf{n}^*_{L_1}(E, \mathbf{r})V(\mathbf{r})\phi_{L_0}(E, \mathbf{r})d\mathbf{r} \\ &= -\alpha\int_{S_A}\left\{\mathbf{n}^*_{L_1}(E, \mathbf{r}), \phi_{L_0}(E, \mathbf{r})\right\}.\, \mathbf{n}dS, \end{aligned} \tag{2.48}$$

and

$$\begin{aligned} s_{L_1L_0} &= -\alpha\lim_{\delta\to 0}\int_{\mathbf{r}\subset V_\delta^A} \mathbf{j}^*_{L_1}(E, \mathbf{r})V(\mathbf{r})\phi_{L_0}(E, \mathbf{r})d\mathbf{r} \\ &= -\alpha\int_{S_A}\left\{\mathbf{j}^*_{L_1}(E, \mathbf{r}), \phi_{L_0}(E, \mathbf{r})\right\}.\, \mathbf{n}dS, \end{aligned} \tag{2.49}$$

with $\mathbf{j}^*_{L_1}(E, \mathbf{r}) = j_{l_1}(\alpha r)Y^*_{L_1}(\hat{\mathbf{r}})$ and $\mathbf{n}^*_{L_1}(E, \mathbf{r}) = n_{l_1}(\alpha r)Y^*_{L_1}(\hat{\mathbf{r}})$. Using the relation between the s-matrix and the t-matrix in equation (2.30), the t-matrix can be written in terms of the sine and cosine matrices

$$\mathbf{t} = -\frac{1}{\alpha}\mathbf{s}(\mathbf{c} - i\mathbf{s})^{-1}. \tag{2.50}$$

At this point, it is useful to insert the definition of the kernel into equation (2.33) to obtain

$$\phi_{L_0}(E, \mathbf{r}) = \sum_L Y_L(\hat{\mathbf{r}})j_l(\alpha r)c_{LL_0}(E, r) - \sum_L Y_L(\hat{\mathbf{r}})n_l(\alpha r)s_{LL_0}(E, r), \tag{2.51}$$

a generalization of the equation used by Calogero [1] in his scattering theory. Unlike Calogero's function, this one makes it possible to deal with non-spherical potentials. The cosine and sine functions

$$c_{L_1L_0}(r) = \delta_{L_1L_0} - \alpha\lim_{\delta\to 0}\int_{|\mathbf{r}'|\subset r} \mathbf{n}^*_{L_1}(E, \mathbf{r}')V(\mathbf{r}')\phi_{L_0}(E, \mathbf{r}')d\mathbf{r}' \tag{2.52}$$

and

$$s_{L_1L_0}(r) = -\alpha\lim_{\delta\to 0}\int_{|\mathbf{r}'|\subset r} \mathbf{j}^*_{L_1}(E, \mathbf{r}')V(\mathbf{r}')\phi_{L_0}(E, \mathbf{r}')d\mathbf{r}' \tag{2.53}$$

become equal to the ones defined above only for $r > R_{bs}$, but, as above, $\phi_{L_0}(E, \mathbf{r})$ is defined for all **r**.

It should be noted that the relationship between the functions $\phi_L(E, \mathbf{r})$ and $\psi_L(E, \mathbf{r})$ expressed in equation (2.40)

$$\psi_L(E, \mathbf{r}) = \sum_{L'} \phi_{L'}(E, \mathbf{r}) b^{-1}_{L'L}, \tag{2.54}$$

constitutes a proof that the scattering wave function produced by the Lippmann–Schwinger equation is regular at the origin.

2.6 Spherical potentials

For a spherical scatterer, $V(\mathbf{r}) = V(r)$, the t-matrix is diagonal in L, and equation (2.28)

$$\psi_l(E, r) = j_l(\alpha r) - i\alpha h_l^+(\alpha r) t_l(E), \tag{2.55}$$

where we have introduced the new function $\psi_l(E, r)$ by

$$\psi_L(E, \mathbf{r}) = Y_L(\mathbf{r})\psi_l(E, r). \tag{2.56}$$

It is also possible to find the solution of the Schrödinger equation for a given l, $\phi_l(E, r)$, by direct solution of the radial equation,

$$\left[-\frac{\hbar^2}{2m}\frac{1}{r^2}\frac{d}{dr}\left(r^2 \frac{d}{dr} \right) + \frac{\hbar^2 l(l+1)}{2mr^2} + V(r) \right] \phi_l(E, r) = E\phi_l(E, r). \tag{2.57}$$

We are only interested in the solution of this equation that is regular at the origin

$$\lim_{r \to 0} \phi_l(E, r) = j_l(\alpha r). \tag{2.58}$$

When r is outside the range of the potential, the radial solution can be written as a linear combination of the spherical Bessel functions that are the solutions of the radial equation with $V(r) = 0$

$$\phi_l(E, r) = j_l(\alpha r) \cos \delta_l - n_l(\alpha r) \sin \delta_l. \tag{2.59}$$

This equation defines the phase shifts $\delta_l(E)$. For spherical potentials, the relation in equation (2.54) takes the form

$$\psi_l(E, r) = e^{i\delta_l} \phi_l(E, r), \tag{2.60}$$

and from this we obtain an expression for the t-matrix in terms of the phase shifts

$$t_l = -\frac{1}{\alpha} e^{i\delta_l} \sin \delta_l. \tag{2.61}$$

To find the phase shifts δ_l, the radial equation is first solved for $\phi_l(E, r)$ using some techniques of numerical integration. Assuming that the potential is zero outside a circle of radius a, the continuity of the wave function and its derivative gives the equations

$$\left(\frac{d\phi_l(r)}{dr}\right)_{r=a} = \alpha A_l \left[\frac{dj_l(x)}{d(x)}\cos\delta_l - \frac{dn_l(x)}{d(x)}\sin\delta_l\right]_{x=\alpha a}. \tag{2.62}$$

This can be looked upon as two equations for the two unknowns A_l and δ_l. They can be solved easily using the identity for Bessel functions

$$j_l(x)\frac{dn_l(x)}{dx} - n_l(x)\frac{dj_l(x)}{dx} = \frac{1}{x^2}. \tag{2.63}$$

The solution is

$$\tan\delta_l = \frac{\left[\alpha\phi_l(r)\frac{dj_l(x)}{dx} - \frac{d\phi_l(r)}{dr}j_l(x)\right]_{x=\alpha r}}{\left[\alpha\phi_l(r)\frac{dn_l(x)}{dx} - \frac{d\phi_l(r)}{dr}n_l(x)\right]_{x=\alpha r}}. \tag{2.64}$$

The process we have outlined for solving the scattering problem for a spherical potential is called the partial wave method.

The method of partial waves is particularly easy for a mathematical model of a rigid sphere that is a spherical square well in which the potential is infinite for $r \leqslant a$. Since $\phi_l(a) = 0$ for this case, it follows that

$$\tan\delta_l = \frac{j_l(ka)}{n_l(ka)}. \tag{2.65}$$

For low energies and long wavelengths,

$$\begin{aligned} j_l(ka) &= \frac{(ka)^l}{1\cdot 3\cdot 5\cdot \ldots(2l+1)} \\ n_l(ka) &= \frac{1\cdot 3\cdot 5\cdot \ldots(2l-1)}{(ka)^{l+1}}, \end{aligned} \tag{2.66}$$

so

$$\lim_{k\to 0}\tan\delta_l \to \frac{(ka)^{(2l+1)}}{(2l+1)!!(2l-1)!!}. \tag{2.67}$$

This equation is important because it turns out that the scattering from any potential that is bounded in space approaches this limit as the energy approaches zero. From this equation and others, it can be seen that the scattering phase shifts generally become smaller as l becomes bigger.

For spherical potentials, equation (2.39) becomes

$$\phi_l(E, r) = \frac{1}{2}\left[f_l^+(E, \mathbf{r})a_l(E) + f_L^-(E, \mathbf{r})b_l(E)\right]. \tag{2.68}$$

For this case, the quantities we have introduced are the same as the Jost functions, $a_l(E) = L_l^-(E)$ and $b_l(E) = L_l^+(E)$.

2.7 Analytical properties of scattering matrices

A useful starting point for deriving the analytical properties of scattering matrices is the well-known continuity equation

$$\frac{\partial \rho}{\partial t} + \nabla \cdot \mathbf{j} = 0, \tag{2.69}$$

with

$$\rho(\mathbf{r}, t) = |\psi(\mathbf{r}, t)|^2, \tag{2.70}$$

and

$$\mathbf{j} = \frac{\hbar}{i2m}[\psi^*(\mathbf{r}, t)\nabla\psi(\mathbf{r}, t) - \{\nabla\psi^*(\mathbf{r}, t)\}\psi(\mathbf{r}, t)]. \tag{2.71}$$

If the wave function has no explicit time dependence, the partial with respect to t is zero and the equation is simply

$$\nabla \cdot \mathbf{j}(\mathbf{r}) = 0. \tag{2.72}$$

We have shown that the solution of the scattering equation for a given energy can be written

$$\psi(\mathbf{r}) = \sum_L c_L \phi_L(\mathbf{r}). \tag{2.73}$$

Integrating equation (2.72) over a sphere with radius A leads to

$$\int \hat{\mathbf{r}} \cdot \mathbf{j}(\mathbf{r}) A^2 d\Omega = 0. \tag{2.74}$$

We use the form for $\phi_L(\mathbf{r})$ in equation (2.51) and note that the integral of the solid angle leads to a δ-function in L. A few manipulations, including the use of equation (2.63), leads to

$$\tilde{\mathbf{c}}^*\mathbf{s} - \tilde{\mathbf{s}}^*\mathbf{c} = 0. \tag{2.75}$$

In arriving at this conclusion, we used the fact that the coefficients c_L are arbitrary.

We now consider the S-matrix defined in equation (2.41). The product of $\mathbf{S}$ with its Hermitian conjugate is

$$\mathbf{S}\mathbf{S}^\dagger = \mathbf{a}\mathbf{b}^{-1}\tilde{\mathbf{b}}^{*-1}\tilde{\mathbf{a}}^*. \tag{2.76}$$

From equations (2.47) and (2.75) it can be seen that

$$\tilde{\mathbf{b}}^*\mathbf{b} = \tilde{\mathbf{c}}^*\mathbf{c} + \tilde{\mathbf{s}}^*\mathbf{s} = \tilde{\mathbf{a}}^*\mathbf{a}. \tag{2.77}$$

From this it follows that **S** is unitary

$$\mathbf{S}^{-1} = \mathbf{S}^{\dagger}. \tag{2.78}$$

Another matrix that can be used to describe scattering is the Wigner reaction matrix [2]

$$\mathfrak{R} = -\frac{1}{\alpha}\mathbf{s}\mathbf{c}^{-1}. \tag{2.79}$$

The t-matrix can be written in terms of this matrix

$$\mathbf{t} = \mathfrak{R}(\mathbf{I} + i\alpha\mathfrak{R})^{-1}, \tag{2.80}$$

and the s-matrix is

$$\mathbf{S} = (\mathbf{I} - i\alpha\mathfrak{R})(\mathbf{I} + i\alpha\mathbf{r}\mathfrak{R})^{-1}. \tag{2.81}$$

For our purposes, the most useful feature of the reaction matrix is that it is Hermitian

$$\mathfrak{R}^{\dagger} = \tilde{\mathfrak{R}}^{*} = \mathfrak{R}, \tag{2.82}$$

which is obvious from equation (2.75).

Later in the development of MST it will be seen that the use of complex energies leads to significant advantages. When we introduce that extension of the theory, we will have to revisit equations like equation (2.75) and change the notation. It is easy to show that the eigenvalues of a unitary matrix have modulus 1, i.e. $|\lambda| = 1$. For a spherical scatterer, as mentioned above, they are usually written in terms of phase shifts $\lambda_l = e^{i2\delta_l}$. From equation (2.77), it is seen that the diagonal elements of **a** and **b** do not have modulus 1, $a_l = \xi_l e^{i2\delta_l}$ and $b_l = \xi_l e^{i2\delta_l}$. It follows that the identification of the **c** and **s** matrices with cosine and sine matrices should not be taken literally because they do not have the usual normalization. Similar considerations apply to the more general case of non-spherical scatterers because

$$\tilde{\mathbf{c}}^{*}\mathbf{c} + \tilde{\mathbf{s}}^{*}\mathbf{s} \neq 1. \tag{2.83}$$

References

[1] Calogero F 1967 *Variable Phase Approach to Potential Scattering* (New York: Academic)

[2] Wigner E P 1946 *Phys. Rev.* **70** 15

IOP Publishing

Multiple Scattering Theory
Electronic structure of solids
J S Faulkner, G Malcolm Stocks and Yang Wang

Chapter 3

Multiple scattering equations

The Lippmann–Schwinger equation

$$| \psi\rangle = (1 + G_{0+}T) \,|\, \phi\rangle, \tag{3.1}$$

with

$$T = V + VG_0^+V + VG_0^+V + \cdots = V(1 - G_0^+V)^{-1} = V(1 + G_0^+T), \tag{3.2}$$

was introduced in chapter 2. Suppose that the potential V is the sum of N potentials

$$V = \sum_{i=1}^{N} v_i. \tag{3.3}$$

The following analysis is only useful if the potentials in the position representation are localized in space, and that the separate potentials do not overlap. That is

$$v_i(\mathbf{r})v_j(\mathbf{r}) = 0 \tag{3.4}$$

for all $\mathbf{r}$ and $i \neq j$. For example, the potentials could describe the scattering from a cluster of atoms with nuclei located at positions $\mathbf{R}_i$.

3.1 Derivation of multiple scattering equations

Define an operator Q_i so that T can be written as a sum

$$T = \sum_{i=1}^{N} Q_i. \tag{3.5}$$

Inserting these expressions for V and T into equation (3.2) leads to

$$T = \sum_i v_i \left(1 + G_0^+ \sum_j Q_j \right) = \sum_i v_i G_0^+ Q_i + \sum_i v_i \left(1 + G_0^+ \sum_{j \neq i} Q_j \right), \tag{3.6}$$

doi:10.1088/2053-2563/aae7d8ch3

so

$$Q_i = v_i G_0^+ Q_i + v_i \left(1 + G_0^+ \sum_{j \neq i} Q_j \right). \tag{3.7}$$

Moving the first term to the left side of the equation and dividing leads to

$$Q_i = (1 - v_i G_0^+)^{-1} v_i \left(1 + G_0^+ \sum_{j \neq i} Q_j \right) = t^i \left(1 + G_0^+ \sum_{j \neq i} Q_j \right), \tag{3.8}$$

where

$$t^i = v_i (1 + G_0^+ t^i) \tag{3.9}$$

is the t-matrix that describes scattering from the *i*th atom. Iterating equation (3.8) leads to a multiple scattering picture in that

$$T = \sum_i t^i + \sum_i t^i G_0^+ \sum_{j \neq i} t^j + \sum_i t^i G_0^+ \sum_{j \neq i} t^j G_0^+ \sum_{k \neq j} t^k + \cdots. \tag{3.10}$$

The first term in this series describes events in which the wave is scattered from each of the atoms individually and then leaves the cluster. The second term describes events in which the wave scatters from one atom, propagates to a second atom where it is scattered again, and then leaves the cluster. The higher order terms are those for which the wave is scattered by increasing numbers of atoms before it gets out of the cluster. The only rule on these multiple scattering events is that the wave cannot return to site i until it has propagated to at least one other site. The events described by equation (3.10) are illustrated in figure 3.1.

In terms of the operators Q_i, the Lippmann–Schwinger equation, equation (3.1), becomes

$$|\psi\rangle = \left(1 + G_0^+ \sum_{j \neq i} Q_j \right) |\phi\rangle + G_0^+ Q_i |\phi\rangle = |\phi_i^{in}\rangle + |\phi_i^{\text{out}}\rangle. \tag{3.11}$$

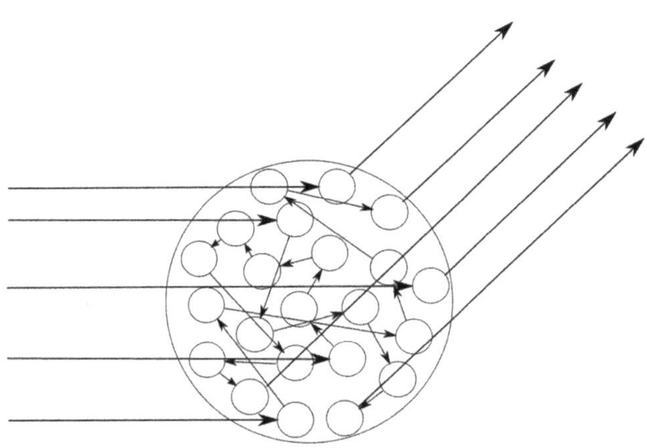

Figure 3.1. An illustration of multiple scattering events.

The site i that we have chosen to focus on can be any of the sites in the cluster. The incoming wave on this site is taken to be

$$\left| \phi_i^{in} \right\rangle = \left(1 + G_{0+} \sum_{j \neq i} Q_j \right) | \phi \rangle = | \phi \rangle + \sum_{j \neq i} \left| \phi_j^{\text{out}} \right\rangle, \tag{3.12}$$

and the outgoing wave from the site is defined as

$$\left| \phi_i^{\text{out}} \right\rangle = G_0^+ Q_i \, | \phi \rangle = G_0^+ t^i \left| \phi_i^{\text{in}} \right\rangle. \tag{3.13}$$

The first of these equations states that the incoming wave on an atomic site is the sum of the wave impinging on the cluster of atoms plus the waves scattered out from all of the other atoms. At first sight, this seems obvious, but one might wonder about the effect of intervening atoms. It is fundamentally a statement about the superposition of solutions of a linear differential equation. The second equation states that the outgoing wave from an atom is obtained from the incoming wave by the solution of a single-site scattering problem.

3.2 Approximations

The lowest level approximation to the scattering matrix for the cluster is

$$T = \sum_i t^i. \tag{3.14}$$

The incoming wave on the cluster can be written

$$\langle \mathbf{r} | \mathbf{k} \rangle = e^{i\mathbf{k}\cdot\mathbf{R}_i} (2\pi)^{-3/2} e^{i\mathbf{k}\cdot\mathbf{r}_i} \tag{3.15}$$

where $\mathbf{r}_i = \mathbf{r} - \mathbf{R}_i$. The Green's function depends only on the separation between $\mathbf{r}$ and $\mathbf{r}'$ so that the origin is immaterial $G_0^+(\mathbf{r} - \mathbf{r}') = G_0^+(\mathbf{r}_i - \mathbf{r}'_i)$. The Lippmann–Schwinger equation in this approximation becomes, in this approximation

$$\psi_{\mathbf{k}}(\mathbf{r}) = \frac{1}{(2\pi)^{3/2}} \left[e^{i\mathbf{k}\cdot\mathbf{r}} + \sum_{i=1}^{N} \frac{e^{ikr_i}}{r_i} f^i(\mathbf{k}, \mathbf{k}'_i) e^{i\mathbf{k}\cdot\mathbf{R}_i} \right], \tag{3.16}$$

where

$$f^i(\mathbf{k}, \mathbf{k}'_i) = -\frac{1}{4\pi} \int e^{-i\mathbf{k}'_i\cdot\mathbf{r}'_i} t^i(\mathbf{r}_i, \mathbf{r}'_i) e^{i\mathbf{k}\cdot\mathbf{r}_i} d\mathbf{r}_i = \frac{(2\pi)^3}{4\pi} \langle \mathbf{k}'_i | \, t^i \, | \, \mathbf{k} \rangle. \tag{3.17}$$

In this formula, the assumption has been made that $r_i \gg r_i'$, and it follows that $\mathbf{k}'_i = k\frac{\mathbf{r}_i}{r_i}$. It can be simplified further by assuming that $r \gg R_i$ for any i, so that

$$\psi_{\mathbf{k}}(\mathbf{r}) = \frac{1}{(2\pi)^{3/2}} \left[e^{i\mathbf{k}\cdot\mathbf{r}} + \frac{e^{ikr}}{r} \sum_{i=1}^{N} f^i(\mathbf{k}, \mathbf{k}') e^{i(\mathbf{k}-\mathbf{k}')\cdot\mathbf{R}_i} \right]. \tag{3.18}$$

This is the form of the multiple scattering equations most frequently used to describe experiments on x-rays scattering from atoms, radio waves scattering from obstacles, and sound waves scattering from bubbles, etc.

As an example, the approximations leading to equation (3.18) are obviously very good for an x-ray diffractometer. The sample size is of the order of millimeters, while the counter is a fraction of a meter away. If the sample is a crystal, the atoms are distributed in a periodic array, and all the $f^i(\mathbf{k}, \mathbf{k}')$ are the same. It is easy to show that when the $\mathbf{R}_i$ form a periodic lattice,

$$\sum_{i=1}^{N} e^{i(\mathbf{k}-\mathbf{k}')\cdot\mathbf{R}_i} = N\sum_{n}\delta(\mathbf{k} - \mathbf{k}' + \mathbf{K}_n), \tag{3.19}$$

where $\mathbf{K}_n$ are the reciprocal lattice vectors. From equations (2.18) and (2.19), it follows that the x-rays are only scattered in directions such that the condition $\mathbf{k} - \mathbf{k}' = \mathbf{K}_n$ is satisfied. This explains the occurrence of Bragg peaks in x-ray diffraction.

As diffraction experiments become more precise, it has become necessary to improve on this simple theory. The first such improvement is attributed to Moliere, who basically went beyond equation (3.14) by including the second term in equation (3.10). This takes into account scattering events in which the incoming particle interacts with two scatterers before leaving the sample. Theories that include pairs or triplets of scatterers are referred to in the literature as multiple scattering theories. This is very different from the present document in which the word 'multiple' refers to an infinite number of scatterers.

3.3 Proof of Korringa's hypothesis

The preceding equations reprise the MST of Kasterin and Ewald. The paradigm shift that led to the use of MST for calculating stationary states in condensed matter was the realization by Korringa that, as the number of scatterers approaches infinity, the multiple scattering equations, equations (3.12) and (3.13), have non-trivial solutions even when the impinging wave $|\phi\rangle$ is zero. Nonzero solutions are found because the scatterings from the atoms within the sample have become self-sustaining.

The purpose of the theory is to find the stationary solutions of the Schrödinger equation, for the case that the potential can be written as a sum of non-overlapping potentials

$$v_{\text{eff}}(\mathbf{r}) = \sum_{n} v_n(\mathbf{r}_n), \tag{3.20}$$

where $\mathbf{r}_n = \mathbf{r} - \mathbf{R}_n$ and, since we are talking about atoms, $\mathbf{R}_n$ is the position of the nth nucleus. We generalize the picture of scatterers in figure 3.1 by assuming that $v_n(\mathbf{r}_n)$ is zero outside a volume Ω_n represented by the boxes in figure 3.2.

In the position representation, the wave function in the cluster $|\psi\rangle = |\phi_i^{\text{in}}\rangle + |\phi_i^{\text{out}}\rangle$ may be found by solving the Schrödinger equation in the neighborhood of any scatterer n. In the preceding chapter, we saw that this function can be expanded in spherical waves

Figure 3.2. A portion of the cluster of scatterers representing atoms in condensed matter. The total number of atoms, N, will increase without bound.

$$\begin{aligned}\psi(E, \mathbf{r}) &= \sum_L \psi_L(E, \mathbf{r}) d_L^n \\ &= \sum_L \left[Y_L(\mathbf{r}_n) j_l(\alpha r_n) - i\alpha \sum_{L'} Y_{L'}(\mathbf{r}_n) h_{l'}^+(\alpha r_n) t_{L',L}^n(E) \right] d_L^n\end{aligned} \tag{3.21}$$

for any $\mathbf{r}$ in the interstitial region where $v_{\text{eff}}(\mathbf{r}) = 0$. The choice of the site n that we choose to expand about is irrelevant. The incoming wave onto and outgoing wave from this site are

$$\psi_n^{\text{in}}(E, \mathbf{r}) = \sum_L Y_L(\mathbf{r}_n) j_l(\alpha r_n) d_L^n, \tag{3.22}$$

and

$$\psi_n^{\text{out}}(E, \mathbf{r}) = -i\alpha \sum_{L,L'} Y_{L'}(\mathbf{r}_n) h_{l'}^+(\alpha r_n) t_{L',L}^n(E) d_L^n. \tag{3.23}$$

Setting $|\phi\rangle = 0$ equation (3.12), the incoming wave on the nth site is the sum of the outgoing waves from all the other sites

$$\sum_L Y_L(\mathbf{r}_n) j_l(\alpha r_n) d_L^n = -i\alpha \sum_{\substack{m=1 \\ m\neq n}}^{N} \sum_{L,L'} Y_{L'}(\mathbf{r}_m) h_{l'}^+(\alpha r_m) t_{L',L}^m(E) d_L^m, \tag{3.24}$$

where we assume the cluster is made up of N atoms. The Hankel function $h_{l'}^{+}(\alpha r_m)$ is singular at $\mathbf{R}_m$, but not in the neighborhood of $\mathbf{R}_n$. It follows that it can be expanded in terms of the complete set of non-singular spherical harmonics

$$-i\alpha Y_{L'}(\mathbf{r}_m)h_{l'}^{+}(\alpha r_m) = \sum_{L} Y_L(\mathbf{r}_n)j_l(\alpha r_n)g_{LL'}^{nm}(E). \tag{3.25}$$

The expansion coefficients, $g_{LL'}^{nm}(E)$, are called free-electron propagators because they describe the movement of the electron from $\mathbf{R}_m$ to $\mathbf{R}_n$ in a region where $v_{\mathrm{eff}}(\mathbf{r}) = 0$. They may be written

$$g_{LL'}^{nm} = -4\pi i\alpha i^{l-l'}\sum_{\Lambda} i^{-\lambda}c_{LL'}^{\Lambda}h_\lambda(\alpha\,|\mathbf{R}_m - \mathbf{R}_n|)\,Y_\Lambda^{*}(\mathbf{R}_m - \mathbf{R}_n), \tag{3.26}$$

where the Gaunt factors

$$c_{LL'}^{\Lambda} = \int Y_\Lambda(\vartheta, \varphi)Y_L^{*}(\vartheta, \varphi)Y_{L'}(\vartheta, \varphi)\sin\vartheta d\vartheta d\varphi, \tag{3.27}$$

are related to the Clebsch–Gordon coefficients.

The meaning of equation (3.24) is that there are stationary wave functions for the crystal if and only if a set of coefficients d_L^n such that

$$\sum_{m=1}^{N}\sum_{L'} Y_L(\mathbf{r}_n)j_l(\alpha r_n)\hat{M}_{LL'}^{nm}(E)d_{L'}^{m}(E) = 0, \tag{3.28}$$

with the matrix $\hat{\mathbf{M}}$ being completely determined by the geometry of the sample and the scattering strength of the atoms

$$\hat{\mathbf{M}}^{nm}(E) = \mathbf{I}\delta_{nm} - (1 - \delta_{nm})\mathbf{g}^{nm}(E)\mathbf{t}^{m}(E). \tag{3.29}$$

Because the spherical Bessel functions and harmonics are a complete set, it follows that the column vector d_L^n must satisfy the set of homogeneous equations

$$\hat{M}_{LL'}^{nm}(E)d_{L'}^{m}(E) = 0, \tag{3.30}$$

for all L, or in matrix notation

$$\hat{\mathbf{M}}(E)\mathbf{d}(E) = 0. \tag{3.31}$$

From elementary math, we know that non-trivial solutions for the d_L^n exist only for the specific energies such that

$$\det\hat{\mathbf{M}}(E_i) = 0. \tag{3.32}$$

Since the sample under consideration is infinitely large, it can be difficult to find these eigenvalues, but the fact that they exist constitutes a proof that Korringa's hypothesis is correct. The remainder of this treatise is devoted to the study of ways to find the eigenvalues and wave functions for various configurations of atoms using this formalism.

The preceding equations have been rewritten in different ways which are mathematically equivalent. For example, the coefficients that define the stationary wave functions can be replaced by

$$c_{L'}^{m}(E) = \sum_{L} t_{L',L}^{m}(E) d_{L}^{m}(E). \tag{3.33}$$

Equation (3.31) is then replaced with

$$\mathbf{M}(E)\mathbf{c}(E) = 0, \tag{3.34}$$

where the matrix $\mathbf{M}$ has elements

$$M_{LL'}^{nm}(E) = m_{LL'}^{n}(E)\delta_{nm} - (1 - \delta_{nm})g_{LL'}^{nm}(E). \tag{3.35}$$

The matrices $\mathbf{m}^n$ are the inverse of the t-matrices

$$\mathbf{m}^n = (\mathbf{t}^n)^{-1} = -\alpha\mathbf{c}^n\mathbf{s}^{n-1} + i\alpha\mathbf{I}, \tag{3.36}$$

and, from equation (3.26). The eigenvalues found by setting the determinant of $\mathbf{M}$ equal to zero are, of course, the same as the ones that satisfy equation (3.32). The frequently cited advantage to this form of the MST equations is that it clearly separates the geometrical aspects of the system contained in the propagators $g_{LL'}^{nm}(E)$ from the part that depends on the particular species of atom on a site $m_{LL'}^{n}(E)$.

After the eigenvalues have been found, the corresponding eigenvectors $c_L^n(E_i)$ can be solved for, and with them the wave function is, from equation (3.24),

$$\psi_i(E_i, \mathbf{r}) = \sum_{m=1}^{N}\sum_{L,L'} \psi_L(E_i, \mathbf{r}_m) m_{LL'}^{m} c_{i,L'}^{m} = \sum_{m=1}^{N}\sum_{L} Z_L^m(E_i, \mathbf{r}_m) c_{i,L}^{m}. \tag{3.37}$$

The physical interpretation of the function

$$Z_L^m(E, \mathbf{r}_m) = \sum_{L'} \psi_{L'}(E, \mathbf{r}_m) m_{L'L}^{m}, \tag{3.38}$$

is not as apparent as that of $\psi_L(E, \mathbf{r}_n)$, but it is more convenient algebraically. Using equation (2.54) it can also be written

$$Z_L^m(E, \mathbf{r}_m) = -\alpha \sum_{L'} \phi_{L'}^{m}(E, \mathbf{r}_m)(\mathbf{s}^m)_{L'L}^{-1}. \tag{3.39}$$

Since $\phi_L^m(E, \mathbf{r}_n)$ and $\psi_L^m(E, \mathbf{r}_n)$ are defined for all $\mathbf{r}$ and are regular at the origin, the same applies to $Z_L^m(E, \mathbf{r}_n)$. When $\mathbf{r}_m$ is outside the range of the potential $v_m(\mathbf{r}_m)$, the function can be written

$$Z_L^m(E, \mathbf{r}_m) = \sum_{L'} Y_{L'}(\mathbf{r}_m) j_{l'}(\alpha r_m) m_{L'L}^{m} - i\alpha Y_L(\mathbf{r}_m) h_l^{+}(\alpha r_m), \tag{3.40}$$

or

$$Z_L^m(E, \mathbf{r}_m) = -\alpha\left[\sum_{L'} Y_{L'}(\hat{\mathbf{r}}_m) j_{l'}(\alpha r_m)(\mathbf{c}\mathbf{s}^{-1})_{L'L}^{m} - Y_L(\hat{\mathbf{r}}_m) n_l(\alpha r_m)\right]. \tag{3.41}$$

3.4 The Korringa–Kohn–Rostoker band theory

The MST equations of the preceding section can be solved most easily if the atoms are on the sites of an infinite periodic lattice because Bloch's theorem can be used to simplify the calculation. This is the case treated by Korringa, Kohn, and Rostoker, and leads to what is called KKR band theory. The matrix $\mathbf{M}(E)$, defined in equation (3.35) may be written

$$\mathbf{M} = \mathbf{m} - \mathbf{g}. \tag{3.42}$$

For a periodic system with one atom per unit cell, $\mathbf{m}$ has the identical block matrices $\mathbf{m}^n$ reproduced down the diagonal. These matrices are made finite in size by ignoring angular momenta greater than $l_{\max}$, so that they have the dimension $(l_{\max} + 1)^2$. The elements of $\mathbf{g}$ are given by equation (3.26) with the nuclear positions $\mathbf{R}_n$ and $\mathbf{R}_m$ on lattice sites.

As explained in elementary solid state physics texts, the lattice vectors may be written

$$\mathbf{R}_i = m_i^1\mathbf{a}_1 + m_i^2\mathbf{a}_2 + m_i^3\mathbf{a}_3, \tag{3.43}$$

where the $\mathbf{a}_i$ are basis vectors and m_i^j are integers. The group theoretical consequence of the invariance of the potential function under translations through $\mathbf{R}_m$ is known as Bloch's theorem

$$\psi(E, \mathbf{r} + \mathbf{R}_m) = e^{i\mathbf{k}\cdot\mathbf{R}_m}\psi(E, \mathbf{r}), \tag{3.44}$$

where the $\mathbf{k}$ are real vectors that exist in a reciprocal space. The geometry of reciprocal space is defined by the lattice vectors

$$\mathbf{K}_i = n_i^1\mathbf{b}_1 + n_i^2\mathbf{b}_2 + n_i^3\mathbf{b}_3, \tag{3.45}$$

where the n_i^j are integers and basis vectors in reciprocal space are defined so that

$$\mathbf{a}_i \cdot \mathbf{b}_j = 2\pi\delta_{ij}. \tag{3.46}$$

It is obvious that

$$e^{i\mathbf{K}_i\cdot\mathbf{R}_j} = 1. \tag{3.47}$$

The unit cell in reciprocal space is called a Brillouin zone. It has the volume

$$\Omega = \mathbf{b}_1 \cdot (\mathbf{b}_2 \times \mathbf{b}_3) = \frac{(2\pi)^3}{V}, \tag{3.48}$$

where V is the volume of a unit cell in real space.

The mathematics of k-space is simplified by invoking periodic boundary conditions on the solutions

$$\psi(E, \mathbf{r} + N_i\mathbf{a}_i) = \psi(E, \mathbf{r}), \tag{3.49}$$

where the N_i are large integers that will eventually approach infinity. There are $N = N_1N_2N_3$ lattice sites in the fundamental cell. The periodic boundary conditions convert the continuous set of Bloch vectors to a discrete set that are written

$$\mathbf{k}_j = k_j^1\mathbf{b}_1 + k_j^2\mathbf{b}_2 + k_j^3\mathbf{b}_3. \tag{3.50}$$

In order for equation (3.49) to be satisfied, the coefficients must be

$$k_j^i = \frac{n_j}{N_i} \quad \text{where } n_j = 0, 1, 2, \ldots, N_i - 1. \tag{3.51}$$

There are N k-vectors in a Brillouin zone.

This simple process of counting the number of states in a Brillouin zone is used to explain properties of solids that were previously inexplicable. According to the Pauli exclusion principle, two electrons can be put into an eigenstate corresponding to a given k-vector, one with spin up and the other with spin down. This means that a metal made of alkali atoms like potassium, that has one conduction electron per atom, will have a half-filled Brillouin zone and a spherical Fermi surface. A metal made of alkali-earth atoms like calcium will have a full Brillouin zone and virtually no free Fermi surface. This explains the differences in the properties of the two metals, such as the low conductivity of calcium compared to potassium.

Consider a unitary matrix $\mathbf{U}$ defined by

$$U_{LL'}^{ij} = \frac{1}{\sqrt{N}} \exp(-i\mathbf{R}_i \cdot \mathbf{k}_j)\delta_{LL'}, \tag{3.52}$$

where $\mathbf{k}_j$ is a vector in the reciprocal lattice. The relations

$$\begin{aligned}(\mathbf{U}\mathbf{U}^\dagger)_{LL'}^{ij} &= \delta_{LL'}\frac{1}{N}\sum_{m=1}^{N} e^{-i(\mathbf{R}_i-\mathbf{R}_j)\cdot\mathbf{k}_m} = \delta'_{ij}\delta_{LL'} \\ (\mathbf{U}^\dagger\mathbf{U})_{LL'}^{ij} &= \delta_{LL'}\frac{1}{N}\sum_{m=1}^{N} e^{i(\mathbf{k}_i-\mathbf{k}_j)\cdot\mathbf{R}_m} = \delta'_{ij}\delta_{LL'},\end{aligned} \tag{3.53}$$

prove that $\mathbf{U}$ is unitary. The prime on the Kronecker delta means that we are ignoring the case that any of the vectors fall outside of the fundamental cells.

We are going to use the U-matrices to transform the N by N matrix $\mathbf{g}$, but there is a problem. In order to carry out the Bloch transformation it is necessary to sum over all of the $\mathbf{g}^{ij} = \mathbf{g}(E, R^i - R^j)$, including those for which $i = j$. From the derivations above, it can be seen that these matrices are excluded from the sums and thus the blocks along the diagonal of $\mathbf{g}$ are zero. From equation (3.26) we get a value for $\mathbf{g}^{nn}$

$$g_{LL'}^{nn} = -4\pi i\alpha i^{l-l'}\sum_{\Lambda} i^{-\lambda} c_{LL'}^{\Lambda} Y_{\Lambda}^{*}(0, 0) = -i\alpha\delta_{LL'}. \tag{3.54}$$

Inserting this quantity into the sum and then subtracting it out leads to

$$\begin{aligned}&(\mathbf{U}^\dagger\mathbf{g}\mathbf{U})^{nm} \\ &= \frac{1}{N}\left\{\sum_{j=1}^{N}\left[\sum_{i=1}^{N} \exp(i\{\mathbf{R}_i - \mathbf{R}_j\} \cdot \mathbf{k}_n) g_{LL'}^{ij}(E) \exp(-i\mathbf{R}_j \cdot \{\mathbf{k}_m - \mathbf{k}_n\})\right] + i\alpha\delta_{LL'}\right\} \\ &= B_{LL'}(E, \mathbf{k}_n)\delta_{nm} + i\alpha\delta_{nm}\delta_{LL'} = g_{LL'}(E, \mathbf{k}_n)\delta_{nm}.\end{aligned} \tag{3.55}$$

This transformation would not work without including $\mathbf{g}^{ii}$ because then the sum over i would depend on j.

The elements of the matrix $\mathbf{B}(E, \mathbf{k}_n)$ are called the KKR structure constants. After the Bloch transformation, the matrix $\mathbf{g}$ is in block-diagonal form, with a matrix of structure constants in each diagonal block. Using equation (3.36), The transform of the matrix $\mathbf{M}$ has the submatrices

$$(\mathbf{U}^{\dagger}\mathbf{M}\mathbf{U})^{nn} = \mathbf{M}(E, \mathbf{k}_n) = -\alpha\mathbf{c}(E)\mathbf{s}(E)^{-1} - \mathbf{B}(E, \mathbf{k}_n), \tag{3.56}$$

down the diagonal, and zeros everywhere else. From the notation, the scattering matrices in each submatrix are the same, while there is a different set of structure constants for each of the $\mathbf{k}_n$ in the Brillouin zone. The submatrices are made finite in size by ignoring the elements corresponding to angular momenta larger than $l_{\max}$. Thus, the $\mathbf{M}(E, \mathbf{k}_n)$ are of dimension $(l_{\max} + 1)^2$. Since there are N k-vectors in the Brillouin zone, the dimension of $\mathbf{U}^{\dagger}\mathbf{M}\mathbf{U}$ is $N \times (l_{\max} + 1)^2$.

The energy eigenvalues corresponding to a given $\mathbf{k}_n$ are the zeros of the determinant of the matrix in equation (3.56), i.e., the eigenvalues $E_j(\mathbf{k}_n)$ are found from

$$\det \mathbf{M}\left(E_j, \mathbf{k}_n\right) = 0. \tag{3.57}$$

The coefficients that define the wave functions are the solutions of

$$\sum_{L'} M_{LL'}\left(E_j, \mathbf{k}_n\right) c_{L'}\left(E_j, \mathbf{k}_n\right) = 0. \tag{3.58}$$

Equations (3.57) and (3.58) are called the KKR equations, and the matrix $\mathbf{M}(E, \mathbf{k})$ is called the KKR matrix. The generalization of this equation for crystals with several atoms in a unit cell is straightforward.

The major contribution of Ewald to MST is the derivation of a mathematically subtle yet powerful technique for calculating the structure constants defined in equation (3.55) [5]. Ham and Segal used the Ewald theorem in their early KKR calculations.

If $N = \infty$ and the scatterers are uniformly distributed throughout space, it is easy to show that the sum on the right side of equation (3.24) should equal infinity. The mathematics is the same as that which leads to Olber's paradox that the light from all of the stars in the Universe should illuminate the sky at night as much as during the day. How can it be reconciled with the fact that Ewald's technique gives a finite sum? The reasons that can be put forward for this being acceptable are analogous to the ones put forward to circumvent Olber's paradox. First, even though samples of condensed matter are large, they are not really infinite. Therefore, it is not correct to include an infinity of incoming waves. Second, the reason we need large values of N is to obtain solutions for bulk states and eliminate the effects of surface states. The Ewald method sums the incoming waves in two ways, a real space sum and a sum in k-space. Both of these sums are truncated in practical calculations. It appears that the surface states are lost in this process, and this is a good thing for the calculations.

3.5 Constant energy surfaces

The derivation of equation (3.57) might imply that it is solved by setting **k** equal to one of its possible values and evaluating the determinant for a series of energies to find the eigenvalues. Another search method is to set E constant and scan k-space for the zeros of the determinant. This is particularly useful for those cases where the primary interest is in the shape of the Fermi surface and the electronic states with energies near the Fermi energy.

An example for which this was particularly useful is the calculation of the Fermi surface of metallic technetium [1]. Technetium has a hexagonal close packed structure and is unique because, although its atomic number is only 43, it has no stable isotopes. It is used in tens of millions of medical diagnostic procedures annually, making it the most commonly used radioisotope. It has a superconducting transition temperature of 7.8°K, which is high for any metal and easily the highest for any hcp metal.

The eigen-energies for all k-vectors in the Brillouin zone can be found from those in the one twenty-fourth of the zone shown in figure 3.3 by applying rotation and reflection operators. In the same figure, we show the intersection of the Fermi surface with the bounding surface of this fundamental segment. The shape of the Fermi surface throughout the Brillouin zone can be pictured by making a three-dimensional model. This can be done by cutting out the shape in the figure and then folding it in an obvious way.

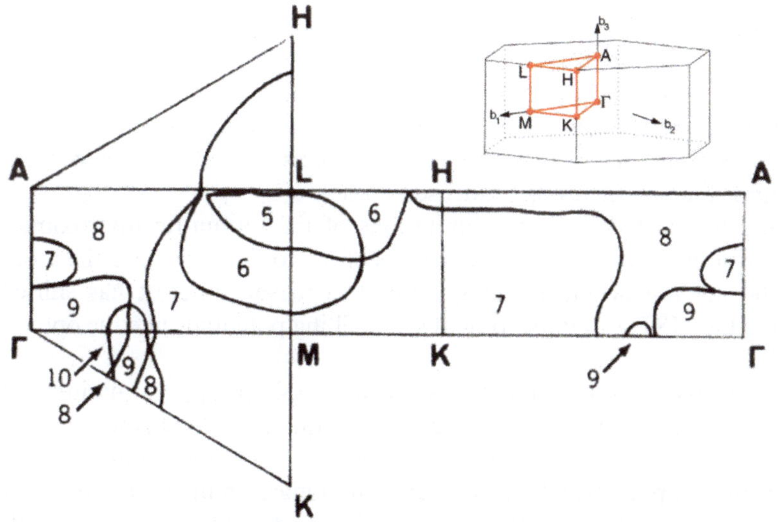

Figure 3.3. Intersection of the Fermi surface of technetium with the symmetry planes that bound the fundamental one twenty-fourth of the Brillouin zone, which is indicated in the appended drawing of the zone. The numbers indicate the number of filled states corresponding to the k-vectors in the various regions of the Brillouin zone. Reprinted with permission from [1] ©1977 American Physical Society.

It is clear from the complexity of this Fermi surface that it would be impossible to determine without using a constant energy search. This was particularly true at the time that these calculations were done.

Another feature of the KKR band theory method is that the eigenfunctions at any energy are automatically orthogonal to the core states. This is different from methods based on the Rayleigh–Ritz variational method, which includes all of the 'linearized' methods. It should be recalled that the variational method only assures accuracy for the lowest eigenvalue in the system, which is the lowest energy core state. The approximations to the higher energy states rely on the assumption that they are orthogonal to the lower energy states. To the degree that this is not true, inaccuracies will be introduced.

3.6 Space-filling potentials

The multiple scattering picture that leads to the KKR equations described above is quite intuitive. It is even more so if we assume that the potentials are spherically symmetric within some radius r_{mt}, and are zero for $r > r_{mt}$. This 'muffin-tin' approximation was the only case considered by Korringa, Kohn, and Rostoker, and it was used in all of the early band theory calculations. The scattering matrices that appear in equation (3.56) are simplified to

$$-\alpha \mathbf{cs}^{-1} = -\alpha \cot \delta_l \delta_{LL'}, \tag{3.59}$$

with the phase shifts calculated from equation (2.36).

The muffin-tin approximation works remarkably well, particularly for metals. Calculations using that approximation explained the highly precise experimental measurements on Fermi surfaces that demonstrated that the one-electron approximation gives a very accurate description of the electronic states in solids. Muffin-tin calculations were the work-horse in the process that revolutionized our understanding of the properties of condensed matter.

During the past decade, the limitations of the muffin-tin approximation have become clearer. In particular, calculations of the total energy and the atomic forces in a solid are not accurate enough. For this reason, interest has shifted toward extending the MST technique to treat space-filling potentials like the one sketched in figure 3.4.

These are also called full potentials because no shape approximation is applied to them. There was considerable debate over the question of whether or not the MST could be generalized to deal with such potentials. It is not as intuitively obvious as it is for muffin-tin potentials because there is no longer an interstitial region with zero potential in which the free-electron waves can propagate. However, mathematical arguments and extensive model calculations have shown [2] that, not only does MST work for space-filling potentials, but also KKR equations (3.57) and (3.58) apply without modification.

The primary difference between calculations with muffin-tin potentials and full potentials is the degree of computational complexity. For muffin-tin potentials, the scattering matrices are diagonal and $l_{\max} = 3$ is usually adequate. For non-spherical

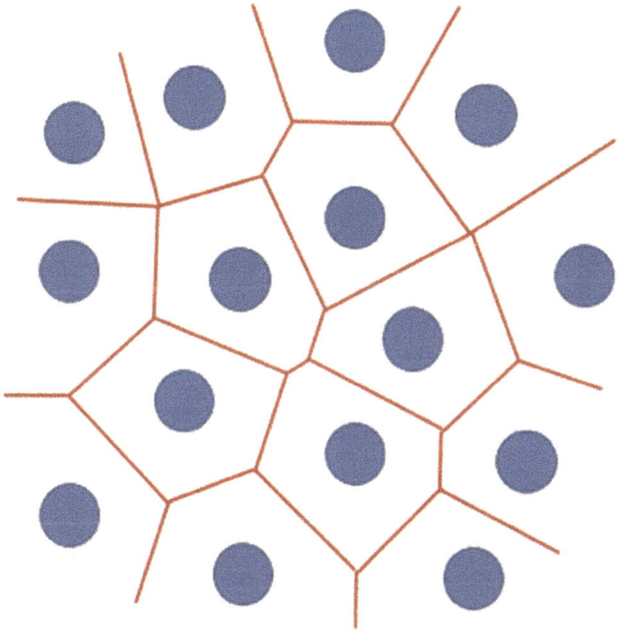

Figure 3.4. A sketch of the boundaries of atomic cells Ω_n for a space-filling potential.

potentials, the scattering matrices and also the step function that describes the shape of the Voronoi polyhedron Ω_n are expanded in l and m, so there can be $(l_{\max} + 1)^4$ terms in the expansions. Experience has shown that it is necessary to keep terms at least as high as $l_{\max} = 8$.

It is also necessary to be extremely careful with the convergence of the computations, both in the solution of the integral equations for the scattering matrices and the l convergence of the matrices. Some calculations that have been the basis of a book suffered from these defects (see reference [27] in chapter 1). After we have developed some more methodology we will show the results of some calculations that eliminated these problems and demonstrated unambiguously that the KKR method works for non-muffin-tin potentials (see reference [28] in chapter 1).

3.7 Pivoted multiple scattering

In muffin-tin calculations, the natural choice is to set the one-electron potential $v_{\text{eff}}(\mathbf{r})$ to zero in the interstitial region. For space-filling potentials, there is no interstitial region and, as is usual in physics, $v_{\text{eff}}(\mathbf{r})$ is fixed only up to an arbitrary constant. There are many ways that the freedom to reset the energy scale can be used in MST calculations. One is to set the energy of the free-electron propagators equal to a constant E_0, so that they become constants $g_{LL'}^{nm}(E_0)$. In band theory calculations, the structure constants depend only on $\mathbf{k}$, $g_{LL'}(E_0, \mathbf{k})$.

Replacing the potential in equation (1.2) by

$$v_{\text{eff}}(\Delta, \mathbf{r}) = \sum_{n}[v_n(\mathbf{r}_n) - \Delta\sigma(\mathbf{r}_n)], \tag{3.60}$$

leads to the original form of the equation if $\Delta = E - E_0$. The function $\sigma(\mathbf{r}_n)$ is the step function that is one inside of Ω_n and zero otherwise. This form of MST is called pivoted multiple scattering theory (PMST). The KKR matrix in PMST is

$$\mathbf{M}(\Delta, \mathbf{k}) = -\alpha_0\mathbf{c}^n(\Delta)\mathbf{s}(\Delta)^{n-1} - \mathbf{B}(E_0, \mathbf{k}). \tag{3.61}$$

In some cases, full potential PMST calculations are computationally faster than those done with ordinary MST. No approximation is being made, so the same results are obtained.

A form of the KKR equation in PMST that yields the same eigenvalues and eigenfunctions is

$$\mathbf{P}(\Delta, \mathbf{k}) = -\alpha_0\mathbf{c}^n(\Delta) - \mathbf{B}(E_0, \mathbf{k})\mathbf{s}(\Delta). \tag{3.62}$$

It is found that the scattering matrices are smoother functions of the displacement of the potential Δ than of the energy E, so they can be expanded in a power series

$$\begin{aligned} \mathbf{c}^n(\Delta, \mathbf{k}) &= \mathbf{c}_0^n(\mathbf{k}) + \mathbf{c}_1^n(\mathbf{k})\Delta + \mathbf{c}_2^n(\mathbf{k})\Delta^2 + \cdots \\ \mathbf{s}^n(\Delta, \mathbf{k}) &= \mathbf{s}_0^n(\mathbf{k}) + \mathbf{s}_1^n(\mathbf{k})\Delta + \mathbf{s}_2^n(\mathbf{k})\Delta^2 + \cdots, \end{aligned} \tag{3.63a}$$

and the KKR matrix becomes

$$\mathbf{P}(\Delta, \mathbf{k}) = \mathbf{P}_0(\mathbf{k}) + \mathbf{P}_1(\mathbf{k})\Delta + \mathbf{P}_2(\mathbf{k})\Delta^2 + \cdots, \tag{3.64}$$

with

$$\mathbf{P}_i(\mathbf{k}) = -\alpha_0\mathbf{c}_i^n - \mathbf{B}(E_0, \mathbf{k})\mathbf{s}_i^n. \tag{3.65}$$

Including only two terms in the expansion of $\mathcal{M}(\Delta, \mathbf{k})$ leads to a linearized KKR theory (LKKR).

$$(\Delta - \boldsymbol{\varepsilon})\mathbf{d} = 0, \tag{3.66}$$

where

$$\boldsymbol{\varepsilon} = -\mathbf{P}_1^{-1}(\mathbf{k})\mathbf{P}_0(\mathbf{k}), \tag{3.67}$$

and $\mathbf{d}$ is a column matrix. The advantage of the LKKR is that the eigenvalues are found by diagonalizing the matrix $\boldsymbol{\varepsilon}$ rather than searching for the zeros of the determinant. The disadvantage is that it gives accurate eigenvalues only in a limited range centered on E_0. The LKKR can be used as a reference point to analyze some of the other linearized band theories that have been proposed [3].

Keeping three terms in equation (3.64) results in a quadratic KKR theory QKKR. The equation may be rewritten

$$[\Delta^2 - \mathbf{D}(\Delta - \boldsymbol{\varepsilon})]\mathbf{d} = 0, \tag{3.68}$$

where

$$\mathbf{D} = -\mathbf{P}_2^{-1}\mathbf{P}_1. \tag{3.69}$$

Using a standard trick, the eigenvalues Δ_i and eigenvectors $\mathbf{d}_i$ can be found by introducing an auxiliary vector $\mathbf{c}$ defined by

$$(\Delta - \varepsilon)\mathbf{d} = \mathbf{c}. \tag{3.70}$$

Then the quadratic eigenvalue equation is replaced by a linear one

$$\begin{pmatrix} \Delta + \varepsilon - \mathbf{D} & \varepsilon^2 \\ -\mathbf{I} & \Delta - \varepsilon \end{pmatrix}\begin{pmatrix} \mathbf{c} \\ \mathbf{d} \end{pmatrix} = 0. \tag{3.71}$$

Of course, the price that must be paid is that the matrix to be diagonalized has twice the dimension of the original KKR matrix. This QKKR has been tested in practice [4], and found to work very well. The range of energies over which the eigenvalues agree with the ones obtained from a full KKR is quite large.

3.8 Wave functions

To this point, the only aspect of the interaction of an electron with a potential $v_n(\mathbf{r})$ that has been made use of is its scattering. To construct charge densities and self-consistent potentials, it is necessary to know the wave function for all $\mathbf{r}$ including inside the volume Ω_n within which the potential is nonzero.

In cluster calculations, the total wave function that satisfies these conditions is

$$\psi(E_F, \mathbf{r}) = \sum_{\substack{i \\ E_i \leqslant E_F}} \sum_{m=1}^{N} \sum_{L} Z_L^m(E_i, \mathbf{r}) c_{i,L}^m, \tag{3.72}$$

where E_F is the highest filled eigenstate. In band theory calculations, it is

$$\psi(E_F, \mathbf{r}) = \frac{1}{(2\pi)^3} \sum_{\substack{i \\ E_i \leqslant E_F}} \int_{\Omega_{BZ}} \sum_{L} Z_L(E_i(\mathbf{k}), \mathbf{r}) c_{i,L}(\mathbf{k}) d\mathbf{k}, \tag{3.73}$$

where the integral is over the Brillouin zone and the sum is over the bands or portions of bands for which the energy is less than the Fermi energy, E_F.

References

[1] Faulkner J S 1977 *Phys. Rev.* B **16** 736
[2] Gonis A and Butler W H 2000 *Multiple Scattering in Solids* (New York: Springer)
[3] Faulkner J S 1979 *Phys. Rev.* B **19** 6186
[4] Nicholson D M and Faulkner J S 1989 *Phys. Rev.* B **39** 8187
[5] Ewald P P 1916 *Ann. Phys., Lpz.* **49** 1

Chapter 4

Green's functions

Since the nineteenth century when the self-trained mathematician George Green began to use the function that bears his name, the primary task of Green's functions has been to convert differential equations with boundary conditions into integral equations. The free-particle Green's function $G_0^+(E, \mathbf{r}, \mathbf{r}')$ was used in chapter 2 to convert the Schrödinger equation into the Lippmann–Schwinger equation. The Green's functions $K_1(E, \mathbf{r}, \mathbf{r}')$ and $K_2(E, \mathbf{r}, \mathbf{r}')$ were used to transform the same differential equation into integral equations for the ancillary functions $\phi_{L_0}(E, \mathbf{r})$ and $f_{L_0}^{\pm}(E, \mathbf{r})$.

The one-electron Green's function plays a more fundamental role in condensed matter theory. It has been found that there are many advantages, both from the conceptual and the computational point of view, to express the solutions of the one-electron equation, equation (1.2), in terms of Green's functions rather than the wave functions and energy eigenvalues that appear in the original Schrödinger formulation.

Suppose the Hamiltonian for the condensed matter system is H. Then the Green's function is a solution of

$$\lim_{\varepsilon\downarrow 0}(E - H + i\varepsilon)G = I, \tag{4.1}$$

or

$$G = \lim_{\varepsilon\downarrow 0}(E - H + i\varepsilon)^{-1}, \tag{4.2}$$

where $H = H_0 + V$. In the position representation, it is written

$$G(E, \mathbf{r}, \mathbf{r}') = \lim_{\varepsilon\downarrow 0}\langle \mathbf{r} \mid (E - H + i\varepsilon)^{-1} \mid \mathbf{r}'\rangle. \tag{4.3}$$

The charge density is obtained by integration

doi:10.1088/2053-2563/aae7d8ch4

$$\rho(\mathbf{r}) = -\frac{2}{\pi}\text{Im}\int_{-\infty}^{E_F} G(E, \mathbf{r}, \mathbf{r})dE \tag{4.4}$$

where the Fermi energy, E_F, is defined by the requirement that the integral of $\rho(\mathbf{r})$ over all space is equal to the total number of electrons, and the '2' accounts for the electron spins. The density of the one-electron energy states is obtained from the equation

$$\rho(E) = -\frac{2}{\pi}\text{Im}\int_{\text{all space}} G(E, \mathbf{r}, \mathbf{r})d\mathbf{r}. \tag{4.5}$$

Kohn and Hohenberg proved the profound theorem that the total energy of a system of electrons depends only on the density of the electrons $\rho(\mathbf{r})$, hence the name density functional theory for their formulation. Kohn and Sham developed a DFT formula for the total energy

$$E_{\text{total}} = \int_{-\infty}^{E_F} E\rho(E)dE - \frac{1}{2}\int\frac{\rho(\mathbf{r})\rho(\mathbf{r}')}{|\mathbf{r} - \mathbf{r}'|}d\mathbf{r}d\mathbf{r}' + E_{xc}[\rho] - \int\frac{\delta E_{xc}[\rho]}{\delta\rho(\mathbf{r})}\rho(\mathbf{r})d\mathbf{r} \tag{4.6}$$

where the exchange correlation functional $E_{xc}[\rho]$ is determined by the particular method used to calculate the effective one-electron potential. This relationship between the charge density and the total energy is the most fundamental reason for focusing on the Green's function. Other reasons will be pointed out later.

4.1 The free-particle Green's functions and its adjoint

Setting $V = 0$ in equations (4.1) and (4.2) leads to the equations for the free-particle Green's function

$$\lim_{\varepsilon\downarrow 0}(E - H_0 + i\varepsilon)G_0 = I, \tag{4.7}$$

or

$$G_0 = \lim_{\varepsilon\downarrow 0}(E - H_0 + i\varepsilon)^{-1}. \tag{4.8}$$

Since H_0 is the kinetic energy operator, in the position representation and using dimensionless units, equation (4.7) becomes

$$(E + \nabla^2)G_0(E, \mathbf{r}, \mathbf{r}') = \delta(\mathbf{r} - \mathbf{r}'). \tag{4.9}$$

In chapter 2, we saw that the solution of this equation that describes outgoing waves may be written

$$G_0^+(E, \mathbf{r}, \mathbf{r}') = -\frac{1}{4\pi}\frac{e^{i\alpha|\mathbf{r}-\mathbf{r}'|}}{|\mathbf{r} - \mathbf{r}'|}. \tag{4.10}$$

There is a standard expansion

$$G_0^+(E, \mathbf{r}, \mathbf{r}') = -i\alpha\sum_{L'} Y_{L'}(\mathbf{r})j_{l'}(\alpha r)h_{l'}^+(\alpha r')Y_{L'}^*(\mathbf{r}'), \tag{4.11}$$

when $r' > r$, and

$$G_0^+(E, \mathbf{r}, \mathbf{r}') = -i\alpha \sum_{L'} Y_{L'}(\mathbf{r}) h_{l'}^+(\alpha r) j_{l'}(\alpha r') Y_{L'}^*(\mathbf{r}') \tag{4.12}$$

when $r' < r$.

An important feature of Green's functions is that they must obey a symmetry rule

$$\overline{F_0^+(E, \mathbf{r}, \mathbf{r}')} = G_0^+(E, \mathbf{r}', \mathbf{r}), \tag{4.13}$$

where the bar on the left side indicates complex conjugation, and $F_0^+(E, \mathbf{r}, \mathbf{r}')$ is the adjoint Green's function. The differential operator in equation (4.9) is formally self-adjoint, so it might be expected that $F_0^+(E, \mathbf{r}, \mathbf{r}')$ would equal $G_0^+(E, \mathbf{r}, \mathbf{r}')$. This is clearly not the case because the complex conjugate of equation (4.10) converts the Green's function for outgoing waves into the one for incoming waves. The problem is resolved by noting that the outgoing-wave boundary condition prevents the operator from being truly self-adjoint, and the reciprocity relation is

$$G_0^+(E, \mathbf{r}, \mathbf{r}') = G_0^+(E, \mathbf{r}', \mathbf{r}). \tag{4.14}$$

This relation will hold for all of the Green's functions that we deal with.

It is not obvious that the Green's function as written in equations (4.11) and (4.12) satisfies the reciprocity relation, although it would be if we used the sum rule for spherical harmonics

$$P_l(\cos\gamma) = \frac{4\pi}{2l+1} \sum_{m=-l}^{l} Y_{lm}^*(\theta', \phi')\, Y_{lm}(\theta, \varphi) \tag{4.15}$$

where γ is the angle between the unit vectors that point in the directions defined by θ, ϕ and θ', ϕ'. For expressions that contain complex spherical harmonics like equations (4.11) and (4.12), it is useful to introduce the notation

$$G(E, \mathbf{r}', \mathbf{r}) = G(E, \mathbf{r}, \mathbf{r}')^{\bullet}, \tag{4.16}$$

where the dot represents an operation in which we take the complex conjugate of the spherical harmonics and leave the rest of the function unchanged. The operation is easy to carry out because the complex conjugate of a spherical harmonic is

$$Y_{l,m}^*(\mathbf{r}) = (-1)^m Y_{l,-m}(\mathbf{r}). \tag{4.17}$$

Throughout this discussion the implicit assumption has been that the energy is real. Later the discussion will focus on Green's functions evaluated for complex energies. This will allow certain energy integrals to be carried out on contours in complex energy space rather than along the real axis, and it will be seen that such integrals are computationally much faster. Perhaps surprisingly, all of the statements that we have made about the reciprocity relation are unchanged when the energy is complex.

4.2 The Green's function for one scatterer

Using the identity $(A - B)^{-1} = A^{-1} + A^{-1}B(A - B)^{-1}$ with the definition in equation (4.2) gives

$$G = G_0 + G_0 V G. \tag{4.18}$$

Iterating the equation leads to

$$G = G_0 + G_0 T G_0. \tag{4.19}$$

Let us assume that the only potential in V is the atomic potential for the nth atom,

$$V(\mathbf{r}) = v_n(\mathbf{r}). \tag{4.20}$$

Then the Green's function we are interested in is

$$G_n = G_0 + G_0 t^n G_0, \tag{4.21}$$

or in the position representation

$$\begin{aligned} G_n(E, \mathbf{r}, \mathbf{r}') = &- i\alpha \sum_{L'} Y_{L'}(\mathbf{r}) j_{l'}(\alpha r) h_{l'}^{+}(\alpha r') Y_{L'}^{*}(\mathbf{r}') \\ &- \alpha^2 \sum_{L} \sum_{L'} Y_L(\mathbf{r}) h_l^{+}(\alpha r) t_{LL'}^{n} h_{l'}^{+}(\alpha r') Y_{L'}^{*}(\mathbf{r}'), \end{aligned} \tag{4.22}$$

when $\mathbf{r}$ and $\mathbf{r}'$ are outside the region in which $v_n(\mathbf{r}) \neq 0$, Ω_n, and $r' > r$. The matrix element in this equation is

$$t_{LL'}^{n}(E) = \iint j_l(\alpha r) Y_L^{*}(\mathbf{r}) \langle \mathbf{r} | \, t(E) | \mathbf{r}' \rangle Y_{L'}(\mathbf{r}') j_l(\alpha r') d\mathbf{r} d\mathbf{r}'. \tag{4.23}$$

This formula can be rearranged into the form

$$\begin{aligned} &G_n(E, \mathbf{r}, \mathbf{r}') = \\ &-i\alpha \sum_{L} \sum_{L'} \left[Y_{L'}(\mathbf{r}_n) j_{l'}(\alpha r_n) m_{L'L}^{n} - i\alpha Y_L(\mathbf{r}_n) h_l^{+}(\alpha r_n) \right] t_{LL'}^{n} h_{l'}^{+}(\alpha r') Y_{L'}^{*}(\mathbf{r}')\text{,} \end{aligned} \tag{4.24}$$

and, from equation (3.40)

$$G_n(E, \mathbf{r}, \mathbf{r}') = -i\alpha \sum_{L} \sum_{L'} Z_L^{n}(E, \mathbf{r}) t_{LL'}^{n} h_{l'}^{+}(\alpha r') Y_{L'}^{*}(\mathbf{r}'). \tag{4.25}$$

As we know, the advantage of this form is that the function $Z_L^n(E, \mathbf{r})$ is still defined when $\mathbf{r} \subset \Omega_n$.

Continuing the discussion of the preceding section, the dot operation on $G_n(E, \mathbf{r}, \mathbf{r}')$ written in the form of equation (4.25) gives

$$G_n(E, \mathbf{r}, \mathbf{r}')^{\bullet} = -i\alpha \sum_{L} \sum_{L'} Y_{L'}(\mathbf{r}') h_{l'}^{+}(\alpha r') \tilde{t}_{L'L}^{n\bullet} Z_L^{n\bullet}(E, \mathbf{r}), \tag{4.26}$$

where the dotted form of $Z_L^n(E, \mathbf{r})$ is

$$Z_L^{n\bullet}(E, \mathbf{r}) = \sum_{L_1} \tilde{m}_{LL_1}^{n\bullet} j_{l_1}(\alpha r) Y_{L_1}^*(\mathbf{r}) - i\alpha h_l^+(\alpha r) Y_L^*(\mathbf{r})$$
$$= -\alpha \sum_{L'} s_{L'L}^{n\bullet -1} \phi_{L'}^{n\bullet}(E, \mathbf{r}). \tag{4.27}$$

We can solve this last equation for $-i\alpha h_l^+(\alpha r) Y_L^*(\mathbf{r})$ and insert it into equation (4.25) to obtain another form for $G_n(E, \mathbf{r}, \mathbf{r}')$ when $r' > r$

$$G_n(E, \mathbf{r}, \mathbf{r}')$$
$$= \sum_L \sum_{L'} Z_L^n(E, \mathbf{r}) t_{LL'}^n Z_{L'}^{n\bullet}(E, \mathbf{r}') - \sum_L \sum_{L_1} Z_L^n(E, \mathbf{r}) t_{LL_1}^n \tilde{m}_{L_1L}^{n\bullet} j_l(\alpha r') Y_L^*(\mathbf{r}'). \tag{4.28}$$

Wang [1] used Wronskian relations to prove a lemma that the sine and cosine matrices satisfy the equation

$$\tilde{\mathbf{c}}^\bullet \mathbf{s} - \tilde{\mathbf{s}}^\bullet \mathbf{c} = 0. \tag{4.29}$$

From Wang's lemma it is immediately obvious that

$$\tilde{\mathbf{m}}^{n\bullet} = -\alpha \tilde{\mathbf{s}}^{n\bullet -1} \tilde{\mathbf{c}}^{n\bullet} + i\alpha \mathbf{I} = -\alpha \mathbf{c}^n \mathbf{s}^{n-1} + i\alpha \mathbf{I} = \mathbf{m}^n, \tag{4.30}$$

and it follows that

$$\tilde{\mathbf{t}}^{n\bullet} = \mathbf{t}^n. \tag{4.31}$$

Using equation (4.30) in equation (4.28) leads to

$$G_n(E, \mathbf{r}, \mathbf{r}') = \sum_L \sum_{L'} Z_L^n(E, \mathbf{r}) t_{LL'}^n Z_{L'}^{n\bullet}(E, \mathbf{r}') - \sum_L Z_L^n(E, \mathbf{r}) j_l(\alpha r') Y_L^*(\mathbf{r}'). \tag{4.32}$$

In this equation, $\mathbf{r}$ can be inside or outside the domain of nonzero potential, but $\mathbf{r}'$ must be such that $|\mathbf{r}'| \not\subset \Omega_n$. The dot operation on $G_n(E, \mathbf{r}, \mathbf{r}')$ gives the reciprocity relation

$$G_n(E, \mathbf{r}, \mathbf{r}')^\bullet = G_n(E, \mathbf{r}', \mathbf{r}), \tag{4.33}$$

if and only if Wang's lemma in the form of equation (4.31) is satisfied.

In order to write the Green's function for one scatterer when both of the points r and r′ are inside the region of the potential, Ω_n, it is necessary to introduce some new functions. They are solutions of the differential equation for all $\mathbf{r}$

$$(E + \nabla^2 - v_n(\mathbf{r})) F_L^n(E, \mathbf{r}) = 0, \tag{4.34}$$

and satisfy the boundary conditions that they become the free-space solutions

$$\lim_{r \to R_{bs}} J_{L_1}^n(E, \mathbf{r}) = Y_{L_1}(\mathbf{r}) j_{l_1}(\alpha r), \tag{4.35}$$

and

$$\lim_{r \to R_{bs}} N_{L_1}^n(E, \mathbf{r}) = Y_{L_1}(\mathbf{r}) n_{l_1}(\alpha r), \tag{4.36}$$

when $\mathbf{r}$ becomes greater than R_{bs}, the radius of a sphere that bounds the region Ω_n. Then the Green's function

$$G_n(E, \mathbf{r}, \mathbf{r}') = \sum_L \sum_{L'} Z_L^n(E, \mathbf{r}) t_{LL'}^n Z_{L'}^{n\bullet}(E, \mathbf{r}') - \sum_L Z_L^n(E, \mathbf{r}) J_L^{n\bullet}(E, \mathbf{r}'), \tag{4.37}$$

is defined for both $\mathbf{r}$ and $\mathbf{r}'$ anywhere in space with condition $\mathbf{r} <= \mathbf{r}'$. If we choose, we can define

$$H_L^{n+}(E, \mathbf{r}) = J_L^n(E, \mathbf{r}) + iN_L^n(E, \mathbf{r}), \tag{4.38}$$

and write

$$Z_L^n(E, \mathbf{r}) = \sum_{L_1} J_{L_1}^n(E, \mathbf{r}) m_{L_1 L}^n - i\alpha H_L^{n+}(E, \mathbf{r}). \tag{4.39}$$

For $r_n > R_{bs}$ this function becomes equal to the one in equation (3.40). The function that is defined for all $\mathbf{r}$ and becomes equal to the one in equation (3.41) is

$$\begin{aligned} Z_L^n(E, \mathbf{r}_m) &= -\alpha \sum_{L'} \phi_{L'}^n(E, \mathbf{r}_n)(\mathbf{s}^n)_{L'L}^{-1} \\ &= -\alpha\left[\sum_{L'} J_{L'}(E, \mathbf{r}_n)(\mathbf{cs}^{-1})_{L'L}^n - N_{L'}(E, \mathbf{r}_n)\right]. \end{aligned} \tag{4.40}$$

Obviously $Z_L^n(E, \mathbf{r})$ is regular at the origin although $H_L^{n+}(E, \mathbf{r}) = J_L^n(E, \mathbf{r}) + iN_L^n(E, \mathbf{r})$, $N_L^n(E, \mathbf{r})$, and $J_L^n(E, \mathbf{r})$ are not.

Let us now define some matrices

$$\mathbf{Z}(E, \mathbf{r}) = \{Z_{L_1}(E, \mathbf{r}), Z_{L_2}(E, \mathbf{r}), \ldots, Z_{L_M}(E, \mathbf{r})\}, \tag{4.41}$$

$$\tilde{\mathbf{Z}}^\bullet(\varepsilon, \mathbf{r}) = \begin{Bmatrix} Z_{L_1}^\bullet(E, \mathbf{r}) \\ Z_{L_2}^\bullet(E, \mathbf{r}) \\ \cdots \\ Z_{L_M}^\bullet(E, \mathbf{r}) \end{Bmatrix}, \tag{4.42}$$

and

$$\tilde{\mathbf{J}}^\bullet(\varepsilon, \mathbf{r}) = \begin{Bmatrix} J_{L_1}^\bullet(E, \mathbf{r}) \\ J_{L_2}^\bullet(E, \mathbf{r}) \\ \cdots \\ J_{L_M}^\bullet(E, \mathbf{r}) \end{Bmatrix}. \tag{4.43}$$

Then equation (4.37) can be rewritten

$$G_n(E, \mathbf{r}, \mathbf{r}') = \mathbf{Z}^n(E, \mathbf{r}) \cdot \mathbf{t}^n \cdot \tilde{\mathbf{Z}}^{n\bullet}(E, \mathbf{r}') - \mathbf{Z}^n(E, \mathbf{r}) \cdot \tilde{\mathbf{J}}^{n\bullet}(E, \mathbf{r}'). \tag{4.44}$$

At this point we reintroduce the function that is actually used in the calculations, $\phi_L(E, \mathbf{r})$. From equation (4.40)

$$\begin{aligned} G_n(E, \mathbf{r}, \mathbf{r}') = {} & \alpha^2 \boldsymbol{\varphi}^n(E, \mathbf{r}) \cdot \mathbf{s}^{n-1} \cdot \mathbf{t}^n \cdot \tilde{\mathbf{s}}^{n\bullet-1} \cdot \tilde{\boldsymbol{\varphi}}^{n\bullet}(E, \mathbf{r}') \\ & + \alpha \boldsymbol{\varphi}^n(E, \mathbf{r}) \cdot \mathbf{s}^{n-1} \cdot \tilde{\mathbf{J}}^{n\bullet}(E, \mathbf{r}'). \end{aligned} \tag{4.45}$$

From equation (2.50), we know that

$$\mathbf{t}^n = -\frac{1}{\alpha}\mathbf{s}^n(\mathbf{c}^n - i\mathbf{s}^n)^{-1}, \tag{4.46}$$

so the preceding equation becomes

$$\begin{aligned} G_n(E, \mathbf{r}, \mathbf{r}') = &-\alpha\boldsymbol{\varphi}^n(E, \mathbf{r}) \cdot [\tilde{\mathbf{s}}^{n\bullet}(\mathbf{c}^n - i\mathbf{s}^n)]^{-1} \cdot \tilde{\boldsymbol{\varphi}}^{n\bullet}(E, \mathbf{r}') \\ &+\alpha\boldsymbol{\varphi}^n(E, \mathbf{r}) \cdot \mathbf{s}^{n-1} \cdot \tilde{\mathbf{J}}^{n\bullet}(E, \mathbf{r}'). \end{aligned} \tag{4.47}$$

Using Wang's lemma, equation (4.29), it is easy to show that

$$(\tilde{\mathbf{c}}^{n\bullet} + i\tilde{\mathbf{s}}^{n\bullet})(\mathbf{c}^n - i\mathbf{s}^n) = \boldsymbol{\Xi}, \tag{4.48}$$

where

$$\boldsymbol{\Xi} = \tilde{\mathbf{c}}^{n\bullet}\mathbf{c}^n + \tilde{\mathbf{s}}^{n\bullet}\mathbf{s}^n. \tag{4.49}$$

Inserting this result into equation (4.46) leads to

$$\mathbf{t}^n = -\frac{1}{\alpha}\mathbf{s}^n\boldsymbol{\Xi}^{-1}[\tilde{\mathbf{c}}^{n\bullet} + i\tilde{\mathbf{s}}^{n\bullet}], \tag{4.50}$$

and the expression for the Green's function becomes

$$\begin{aligned} G_n(E, \mathbf{r}, \mathbf{r}') = &- i\alpha\boldsymbol{\varphi}^n(E, \mathbf{r}) \cdot \boldsymbol{\Xi}^{-1} \cdot \tilde{\boldsymbol{\varphi}}^{n\bullet}(E, \mathbf{r}') \\ &- \alpha\boldsymbol{\varphi}^n(E, \mathbf{r}) \cdot \boldsymbol{\Xi}^{-1} \cdot \tilde{\mathbf{c}}^{n\bullet} \cdot \tilde{\mathbf{s}}^{n\bullet-1} \cdot \tilde{\boldsymbol{\varphi}}^{n\bullet}(E, \mathbf{r}') \\ &+ \alpha\boldsymbol{\varphi}^n(E, \mathbf{r}) \cdot \mathbf{s}^{n-1} \cdot \tilde{\mathbf{J}}^{n\bullet}(E, \mathbf{r}'). \end{aligned} \tag{4.51}$$

Writing out this equation without the matrix notation gives

$$\begin{aligned} G_n(E, \mathbf{r}, \mathbf{r}') = &-i\alpha\sum_L\sum_L \phi_L^n(E, \mathbf{r})\Xi_{LL'}^{-1}\phi_{L'}^{n\bullet}(E, \mathbf{r}') \\ &-\alpha\sum_L\sum_{L'}\sum_{L_1}\sum_{L_2}\phi_L^n(E, \mathbf{r})\Xi_{LL_1}^{-1}\tilde{c}_{L_1L_2}^{n\bullet}\tilde{s}_{L_2L'}^{n\bullet-1}\phi_{L'}^{n\bullet}(E, \mathbf{r}') \\ &+\alpha\sum_L\sum_{L'}\phi_L^n(E, \mathbf{r})s_{LL'}^{n-1}J_{L'}^{n\bullet}(E, \mathbf{r}). \end{aligned} \tag{4.52}$$

For real energies, this neatly divides the Green's function into a real and imaginary part. Although the last two terms are real, they have poles on the real axis when $\det \mathbf{s}^n = 0$, which means that they could contribute to the imaginary part of the Green's function. The reason that they do not is that they cancel each other.

As pointed out in the preceding chapter, the matrices that appear in these equations are made finite by ignoring matrix elements that correspond to angular momenta higher than some chosen value $l_{\max}$. Then the dimensions of $\mathbf{c}^n$, $\mathbf{s}^n$, $\mathbf{t}^n$, or $\mathbf{m}^n$ are $(l_{\max} + 1)^2$. The relation between $Y_L(\mathbf{r})$ and $Y_L^*(\mathbf{r})$ in equation (4.17) means that the transpose of the dot of any of these matrices can be found by simply rearranging. For example,

$$\tilde{c}^{\bullet}_{l_1,m_1,l_2,m_2} = (-1)^{m_1+m_2}c_{l_2,-m_2,l_1,-m_1}. \tag{4.53}$$

Of course, it is not necessary to do anything to find $\tilde{\mathbf{m}}^{n\bullet}$ or $\tilde{\mathbf{t}}^{n\bullet}$.

We have derived many different forms for the Green's function for one scatterer, equations (4.22), (4.25), (4.32), (4.37), (4.51), or (4.52). Each of these forms will turn out to be useful in a particular application.

4.3 The Green's function for N scatterers

The multiple scattering equations look somewhat different for Green's functions than for the wave functions treated in chapter 3. Let us assume that the potential in the equations

$$G = G_0 + G_0VG = G_0 + G_0TG_0 \tag{4.54}$$

is the sum of N atomic potentials

$$V = \sum_{i=1}^{N} v_i. \tag{4.55}$$

As stated before, the potentials must be non-overlapping $v_i(\mathbf{r})v_j(\mathbf{r}) = 0$. The operators Q_i are written here in terms of scattering path operators τ^{ij}

$$Q_i = \sum_{j=1}^{N} \tau^{ij}. \tag{4.56}$$

Then the total t-matrix is

$$T = \sum_{i=1}^{N}\sum_{j=1}^{N} \tau^{ij}, \tag{4.57}$$

and,

$$\tau^{ij} = t^i\delta_{ij} + t^iG_0\sum_{k\neq i}^{N} \tau^{kj}, \tag{4.58}$$

with the one-atom t-matrix

$$t^i = v_i(1 + G_0^+t^i). \tag{4.59}$$

An equally good equation for the scattering path operator is

$$\tau^{kj} = t^k\delta_{jk} + \sum_{l\neq j=1}^{N} \tau^{kl}G_0t^j. \tag{4.60}$$

When evaluating the Green's function in the position representation $G(E, \mathbf{r}, \mathbf{r}')$, it is usually advantageous to have the points $\mathbf{r}$ and $\mathbf{r}'$ in one cell, say Ω_n. In preparation for this, we rearrange the formula for G in the following manner

$$G = G_n + G_nT^{nn}G_n, \tag{4.61}$$

where

$$G_n = G_0 + G_0 t^n G_0 \tag{4.62}$$

is the one scatterer Green's function investigated in the preceding section, and

$$T^{nn} = \sum_{i \neq n = 1}^{N} \sum_{j \neq n = 1}^{N} \tau^{ij}. \tag{4.63}$$

When we transform equation (4.61) into the position representation, we use equation (4.37) for the first G_n and equations (4.25) and (4.26) for the G_n's in the second term. We use equations (3.23)–(3.25) to convert outgoing waves from one site into incoming waves on the next, so that we arrive at

$$G(E, \mathbf{r}, \mathbf{r}') = \mathbf{Z}^n(E, \mathbf{r}) \cdot \boldsymbol{\tau}^{nn}(E) \cdot \tilde{\mathbf{Z}}^{n\bullet}(E, \mathbf{r}') - \mathbf{Z}^n(E, \mathbf{r}) \cdot \tilde{\mathbf{J}}^{n\bullet}(E, \mathbf{r}'), \tag{4.64}$$

where

$$\tau^{ij}_{LL'}(E) = t^i_{LL'}\delta_{ij} + \sum_{k \neq i = 1}^{N} \sum_{L_1} \sum_{L_2} t^i_{LL_1} g^{ik}_{L_1 L_2} \tau^{kj}_{L_2 L'}, \tag{4.65}$$

and the propagators $g^{ik}_{LL'}$ that take the electron from site i to site k are the ones in equation (3.26).

Let us define the matrices

$$\boldsymbol{\tau} = \begin{pmatrix} \boldsymbol{\tau}^{11} & \boldsymbol{\tau}^{12} & \cdots & \boldsymbol{\tau}^{1N} \\ \boldsymbol{\tau}^{21} & \boldsymbol{\tau}^{22} & & \boldsymbol{\tau}^{2N} \\ \vdots & & \ddots & \vdots \\ \boldsymbol{\tau}^{N1} & \boldsymbol{\tau}^{N2} & \cdots & \boldsymbol{\tau}^{NN} \end{pmatrix}, \tag{4.66}$$

$$\mathbf{g} = \begin{pmatrix} 0 & \mathbf{g}^{12} & \cdots & \mathbf{g}^{1N} \\ \mathbf{g}^{21} & 0 & & \mathbf{g}^{2N} \\ \vdots & & \ddots & \vdots \\ \mathbf{g}^{N1} & \mathbf{g}^{N2} & \cdots & 0 \end{pmatrix}, \tag{4.67}$$

and

$$\mathbf{t} = \begin{pmatrix} \mathbf{t}^1 & 0 & \cdots & 0 \\ 0 & \mathbf{t}^2 & & 0 \\ \vdots & & \ddots & \vdots \\ 0 & 0 & \cdots & \mathbf{t}^N \end{pmatrix}. \tag{4.68}$$

The elements are themselves matrices with angular momentum indices. As in the discussion of the multiple scattering equations in the previous chapter, they are made finite by ignoring terms that correspond to angular momenta higher than some $l_{\max}$, making them of dimension $(l_{\max} + 1)^2$. Then equation (4.65) can be written

$$\boldsymbol{\tau} = \mathbf{t} + \mathbf{t}\mathbf{g}\boldsymbol{\tau}. \tag{4.69}$$

This matrix can easily be solved by taking the inverse of the matrix **M**

$$\boldsymbol{\tau} = \mathbf{M}^{-1}, \tag{4.70}$$

where

$$\mathbf{M} = \begin{pmatrix} \mathbf{m}^1 & -\mathbf{g}^{12} & \cdots & -\mathbf{g}^{1N} \\ -\mathbf{g}^{21} & \mathbf{m}^2 & & -\mathbf{g}^{2N} \\ \vdots & & \ddots & \vdots \\ -\mathbf{g}^{N1} & -\mathbf{g}^{N2} & \cdots & \mathbf{m}^N \end{pmatrix}. \tag{4.71}$$

As before, $\mathbf{m}^i$ is the inverse of the t-matrix for site i.

From this discussion it is clear that in order to find the matrix $\boldsymbol{\tau}^{nn}$ for equation (4.64) we must calculate the elements $g^{ij}_{LL'}$ and $m^{i}_{LL'}$. We use them to create $\underline{\mathbf{M}}$, take the inverse, and then project out the nn block.

The Green's function looks different in the position representation if **r** is in cell Ω_n and $\mathbf{r}'$ is in cell Ω_m

$$G(E, \mathbf{r}, \mathbf{r}') = \mathbf{Z}^n(E, \mathbf{r}) \cdot \boldsymbol{\tau}^{nm} \cdot \tilde{\mathbf{Z}}^{m\bullet}(E, \mathbf{r}'). \tag{4.72}$$

The matrix $\boldsymbol{\tau}^{nm}$ is obtained exactly as before except that, after obtaining the inverse of $\underline{\mathbf{M}}$, the nm block is projected out.

4.4 The Green's function for an infinite periodic lattice

Obviously the process of creating the matrix **M** and taking its inverse is impossible for a very large number of atoms, so the method described in the previous section is only a solution in principle. There are approximate methods for solving the equations, and these lead to the locally self-consistent multiple scattering theory (LSMS) that will be described later. In the preceding chapter, it was shown that the MST calculations are significantly simplified for the case when the nuclei are on the sites of an ordered lattice and all of the atoms are the same. The same methods that were used to obtain the KKR band theory equations can be used to find the Green's function for a periodic lattice.

Periodic boundary conditions are used with the fundamental cell containing N atoms. The transformation matrix with elements $U^{ij}_{LL'}$ that was introduced in equation (3.52) is used to convert **M** into a block diagonal form with KKR matrices $\mathbf{M}(E, \mathbf{k}_i)$ corresponding to each k-vector in the Brillouin zone down the diagonal. Rather than employing these KKR matrices to find the eigenvalues and eigenfunctions as in equations (3.57) and (3.58), we take advantage of their relatively small size to take the inverse of the transformed matrix. We then transform back, so that equation (4.70) becomes

$$\boldsymbol{\tau} = \mathbf{U}(\mathbf{U}^{\dagger}\mathbf{M}\mathbf{U})^{-1}\mathbf{U}^{\dagger}. \tag{4.73}$$

In more detail, this equation is

$$\boldsymbol{\tau}^{mn}(E) = \frac{1}{N}\sum_{i=1}^{N} e^{-i\mathbf{k}_i\cdot(\mathbf{R}_m-\mathbf{R}_n)}[\mathbf{m}(E) - \mathbf{g}(E, \mathbf{k}_i)]^{-1}$$
$$= \frac{\Omega}{(2\pi)^3}\int e^{-i\mathbf{k}\cdot(\mathbf{R}_m-\mathbf{R}_n)}[\mathbf{m}(E) - \mathbf{g}(E, \mathbf{k})]^{-1}d\mathbf{k}. \tag{4.74}$$

where the elements of the matrix $\mathbf{g}(E, \mathbf{k})$ are defined in equation (3.55). The integral is over the Brillouin zone, which has a volume $(2\pi)^3/\Omega$.

Another way of writing the above is in terms of the KKR matrix

$$\boldsymbol{\tau}^{mn}(E) = \frac{\Omega}{(2\pi)^3}\int e^{-i\mathbf{k}\cdot(\mathbf{R}_m-\mathbf{R}_n)}\mathbf{M}(E, \mathbf{k})^{-1}d\mathbf{k}. \tag{4.75}$$

As in the band theory case, the KKR matrix can be written as in equation (3.56), and the Ewald sum can be used to evaluate the structure constants. This leads to calculations in which only matrices of dimension $(l_{\max} + 1)^2$ must be considered.

The most important of the path integrals is

$$\boldsymbol{\tau}^{00}(E) = \frac{\Omega}{(2\pi)^3}\int \mathbf{M}(E, \mathbf{k})^{-1}d\mathbf{k}, \tag{4.76}$$

since the charge density in the central cell, which is of course the same in all cells, is

$$\rho(\mathbf{r}) = \frac{1}{\Omega}\,\mathrm{Im}\left[\int_{E_B}^{E_F} \mathbf{Z}^0(E, \mathbf{r})\cdot\boldsymbol{\tau}^{00}\cdot\tilde{\mathbf{Z}}^{0\bullet}(E, \mathbf{r})dE\right], \tag{4.77}$$

where E_B is the bottom of the conduction band. The density of states per unit volume is

$$\rho(E) = \frac{1}{\Omega}\,\mathrm{Im}\left[\int_{\Omega} \mathbf{Z}^0(E, \mathbf{r})\cdot\boldsymbol{\tau}^{00}\cdot\tilde{\mathbf{Z}}^{0\bullet}(E, \mathbf{r})d\mathbf{r}\right]. \tag{4.78}$$

As usual, these formulae become much simpler for the case of a muffin-tin potential. In equation (4.77), the integral

$$\frac{1}{\Omega}\,\mathrm{Im}\left[\int_{E_B}^{E_F} \mathbf{Z}^0(E, \mathbf{r})\cdot\tilde{\mathbf{J}}^{0\bullet}(E, \mathbf{r})dE\right], \tag{4.79}$$

is left out because the integrand is real for real energies.

4.5 The use of a complex energy

The formulae in the preceding section are correct, but are difficult to apply in actual calculations. For infinite systems, $\boldsymbol{\tau}^{00}(E)$ and $\rho(E)$ are such rapidly varying functions of E that it is virtually impossible to deal with them. This problem can be solved by invoking the fact that when the energy is considered to be a complex variable in equation (4.2), the operator $G(E)$ is an analytic function of E. Cauchy's integral theorem then guarantees that the integral in equation (4.4) will give the same result if it is carried out along the real axis along any other contour in the complex energy plane that starts at E_B and ends at E_F. The function $G(E, \mathbf{r}, \mathbf{r}')$ is much smoother on an energy contour that is well away from the real axis, so numerical integrals like the one under consideration can be done very accurately with only a few points.

The price that must be paid for using a complex contour is that the imaginary part of the integral in equation (4.79) is no longer zero, and it must be calculated accurately. The functions $Z_L^0(E, \mathbf{r})$ are regular at the origin, and are already calculated for the evaluation of equation (4.77). The functions $J_L^0(E, \mathbf{r})$ are not regular at the origin and a complicated inward integration is required to evaluate them. It turns out that these integrations can be accomplished for the case of muffin-tin potentials, so this method is used for complex as well as real energies. As will be discussed in the next section, a different approach must be used when the potentials do not have the muffin-tin form.

4.6 Full potential calculations

The observation that makes it possible to calculate accurate charge densities when the potential is not MT is that the ZJ integral in equation (4.79) has nothing to do with the multiple scattering aspect of the problem, as it is part of the single scatterer Green's function equation (4.44). In principle, we could calculate this integral over the contour in the complex energy plane

$$\rho_C(\mathbf{r}) = \mathrm{Im}\int_C \sum_L \sum_{L'} Z_L^n(z, \mathbf{r})[\tau_{LL'}^{nn} - t_{LL'}^n] Z_{L'}^{n\bullet}(z, \mathbf{r}) dz \tag{4.80}$$

and then get the remainder of the charge $\rho_1(\mathbf{r}) = \rho(\mathbf{r}) - \rho_C(\mathbf{r})$ by integrating the single scatterer Green's function over the real axis. In our analysis of equation (4.52), we concluded that only the first term contributes to $\rho_1(\mathbf{r})$,

$$\rho_1(\mathbf{r}) = \int_{E_B}^{E_A} \sqrt{E} \sum_L \sum_{L'} \phi_L^n(E, \mathbf{r}) \Xi_{LL'}^{-1} \phi_{L'}^{n\bullet}(E, \mathbf{r}) dE. \tag{4.81}$$

Therefore, the problem would seem to be solved.

One of the authors of [2], G M Stocks, made an interesting observation. The ZJ integral, the integral over the real axis of the last term in equation (4.52), does not have to be zero. It is possible for the determinant of the sine matrix $\mathbf{s}^0$ to be zero for some real energy value, E_p. Using Cauchy's theorem, it can be shown that the pole on the real axis, as sketched in the following figure, will result in a nonzero integral. The author also noted that $\rho_1(\mathbf{r})$ is given by equation (4.81), even when there is a pole on the real axis. There is only one explanation for these observations, although it is a surprising one. The contribution from the second term in equation (4.52), which also must be singular at E_p, exactly cancels the contribution from the ZJ integral. This leads to the following expression for $\rho(\mathbf{r})$,

$$\begin{aligned}\rho(\mathbf{r}) = {} & \mathrm{Im}\int_C \sum_L \sum_{L'} Z_L^n(z, \mathbf{r}) \tau_{LL'}^{nn}(z) Z_{L'}^{n\bullet}(z, \mathbf{r}) dz \\ & - \mathrm{Im}\int_c \sqrt{z} \sum_L \sum_{L'} \sum_{L_1} \sum_{L_2} \phi_L^n(z, \mathbf{r}) \Xi_{LL_1}^{-1} \tilde{c}_{L_1L_2}^{n\bullet} \tilde{s}_{L_2L'}^{n\bullet-1} \phi_{L'}^{n\bullet}(z, \mathbf{r}) dz,\end{aligned} \tag{4.82}$$

where both integrals are over the large contour in figure 4.1. Due to the analyticity of the last term, the contour can be shrunk down to the small contour c for that integral.

The advantage in the two methods for calculating $\rho(\mathbf{r})$ that are outlined above is that at no point is it necessary to deal with the difficult function $J_L^n(E, \mathbf{r})$. The question of which is the better formula depends on the atoms that make up the solid. Calculations have been done that validate the preceding algebra and show how it works out for real systems.

We define the following contributions to the total charge on an atom:

$$A_1 = \mathrm{Im}\int_{\Omega_n}\int_C \sum_L \sum_{L'} Z_L^n(z, \mathbf{r})\tau_{LL'}^{nn} Z_{L'}^{n\bullet}(z, \mathbf{r})dzd\mathbf{r} \tag{4.83}$$

$$A_2 = -\mathrm{Im}\int_{\Omega_n}\int_C \sum_L Z_L^n(z, \mathbf{r})J_L^{n\bullet}(z, \mathbf{r})dzd\mathbf{r} \tag{4.84}$$

$$A_3 = \mathrm{Im}\int_{\Omega_n}\int_C \sum_L \sum_{L'} Z_L^n(z, \mathbf{r})t_{LL'}^{n} Z_{L'}^{n\bullet}(z, \mathbf{r})dzd\mathbf{r} \tag{4.85}$$

$$B_1 = \int_{\Omega_n}\int_{E_B}^{E_A} \sqrt{E}\sum_L \sum_{L'} \phi_L^n(E, \mathbf{r})\Xi_{LL'}^{-1}\phi_{L'}^{n\bullet}(E, \mathbf{r}')dEd\mathbf{r} \tag{4.86}$$

$$B_2 = \mathrm{Im}\int_{\Omega_n}\int_c \sqrt{z}\sum_L \sum_{L'} \sum_{L_1} \sum_{L_2} \phi_L^n(z, \mathbf{r})\Xi_{LL_1}^{-1}\tilde{c}_{L_1L_2}^{n\bullet}\tilde{s}_{L_2L'}^{n\bullet-1}\phi_{L'}^{n\bullet}(z, \mathbf{r})dzd\mathbf{r}. \tag{4.87}$$

In table 4.1 we show the results of the calculations of these contributions to the total charge in the conduction band for metallic copper and molybdenum that were

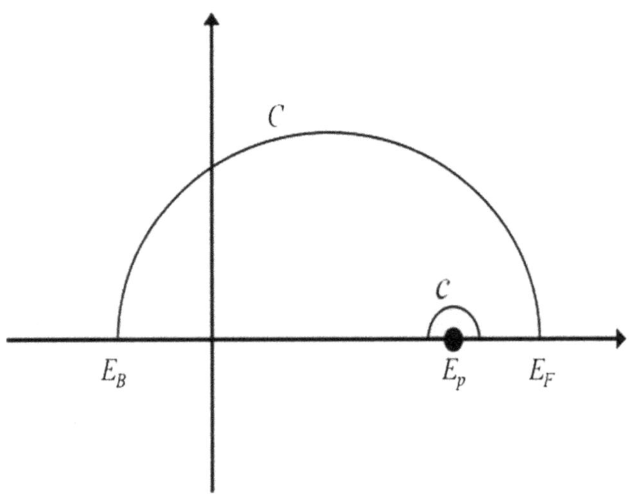

Figure 4.1. Various portions of the Green's function must be integrated from the bottom of the conduction band, E_B, to the Fermi energy, E_F. This integration can be carried out along the real axis or over the contour C in the complex energy plane that connects E_B to E_F. Integrals over the small contour c, which encloses the pole of the inverse of the sine matrix, E_p, are also useful.

Table 4.1. Contributions to the total charge in the conduction band for copper and molybdenum. The unit of charge is e, the charge of an electron.

species	A_1	A_2	A_3	B_1	B_2
Cu (MT)	19.97178	−8.97178	19.60186	10.63008	8.97178
Cu (FP)	19.59501	na	19.22930	10.63428	8.59501
Mo (MT)	6.00000	0.0000	6.36651	6.36651	0.00000
Mo (FP)	6.00000	na	6.38959	6.38959	0.00000

carried out with codes based on the real-space locally self-consistent multiple scattering (LSMS) method. Muffin-tin (MT) and non-muffin-tin full potential (FP) calculations were done on both materials. More conventional KKR band theory methods could have been used, but, for the present purposes, there would be no functional difference in the results. The only significant approximation is the truncation of angular momentum expansions. The scattering matrices are truncated at $\ell_{\max} = 4$, which means that the potentials and charge densities are expanded to $\ell_{\max} = 8$. These contributions are a compact way of demonstrating the relations derived in the preceding section. The contribution A_2 is not evaluated for the FP case since that would require an approximate calculation of $J_L^n(E, \mathbf{r})$.

The MT calculations on Cu in the first row of table 4.1 show that the contribution from ZJ term in equation (4.79) is the negative of the contribution from the second term in equation (4.52), which is a confirmation of the algebra described above. The total charge from the integration of $\rho(\mathbf{r})$ in equation (4.82) is A_1 minus B_2. From rows one and two, this number is seen to be eleven, which is the correct answer for Cu. The contribution obtained by integrating $\rho_C(\mathbf{r})$ over the cell is A_1 minus A_3, while the contribution from $\rho_1(\mathbf{r})$ is B_1. From both rows one and two, we see that $A_1 - A_3 + B_1$ is eleven, as it should be.

The band structure and Fermi surface of molybdenum is known to be more complex than that of copper because the Fermi energy falls in the middle of the d-bands. However, there is no energy between E_B and E_F for which $\det \mathbf{s}^n = 0$. It follows that the ZJ integral is zero for both the FP and MT case, as can be seen from the third and fourth rows in table 4.1.

The charge density for the FP case can be calculated accurately using equations (4.80), (4.81), and (4.82). Using these calculations, it is shown that previous efforts at FP MST calculations (references [26] and [27] in chapter 1) suffered because the approximations used led to a charge density that is inaccurate for small values of r. It is explained in reference [2] that the charge density is quite intricate in that region because the wave functions have to be orthogonal to the core wave functions.

In connection with a study that will be described in chapter 6, one of the authors, Yang Wang [1], discovered another objection to the use of equation (4.81) for calculating the one site contribution to the charge density. The matrix Ξ contains the square of the cosine matrix. From the definition in equation (2.48), it can be seen that $c_{L_1L_0}(E)$ becomes very large when $L_1 \neq L_0$ and L_1 is large because of the behavior of the Bessel function $n_{L_1}(E, r)$. This behavior is exacerbated by the squaring of the matrix, and leads to the result that the elements that are farthest from the diagonal are unreliable even when calculated in double precision. The solution to this computational problem is to use the first term in equation (4.47) to calculate $\rho_1(\mathbf{r})$

$$\rho_1(\mathbf{r}) = \mathrm{Im}\int_{E_B}^{E_A} \sqrt{E}\sum_{L}\sum_{L'}\phi_L^n(E, \mathbf{r})\mathbf{X}_{LL'}^{-1}\phi_{L'}^{n\bullet}(E, \mathbf{r}')dE, \tag{4.88}$$

where

$$\mathbf{X} = \tilde{\mathbf{s}}^{n\bullet}(\mathbf{c}^n - i\mathbf{s}^n). \tag{4.89}$$

Unlike $\boldsymbol{\Xi}$, the matrix $\mathbf{X}$ is complex, even when evaluated on the real axis. However, it is found that the calculations are much more convergent if that complex quantity is calculated and then the imaginary part is taken. Calculations of the total charge contribution from the single site have been carried out using both methods on copper and molybdenum. There is no perceptible error when $\ell_{\max}$ is 2 or 3, but the difference is in the fourth decimal place when $\ell_{\max}$ is 4 or 5. This may not seem like a big error, but, on the scale of the calculations being contemplated, it is significant.

The next step in the evolution of numerical methods for full potential MST calculations was published in connection with a study on relativistic MST to be described in chapter 8 [3]. It applies equally well to nonrelativistic calculations. The proposal is to study the poles in the matrix $\mathbf{X}^{-1}$, or the zeros of $\mathbf{X}$. The most important poles that have $\mathrm{Re}\, E \leqslant 0$ fall on the real axis. Those for which $\mathrm{Re}\, E \geqslant 0$ fall below the real axis and have $\mathrm{Im}E \leqslant 0$. The authors of [3] demonstrate that knowledge of these poles makes it much easier to do FP calculations with great accuracy.

To sum up, before the publication of Rusanu *et al* in 2011, a truly accurate full potential MST calculation had never been done. There are now four different methods to do this that have been tested. The first makes use of equations (4.80) and (4.81). This is the most straightforward technique from a mathematical point of view. It's limitations have been pointed out, but it would be adequate for many applications. The second method, based on equation (4.82), is more complex theoretically, but it has been used with success. The third method corrects the limitations of the first by replacing equation (4.81) with equation (4.88). The fourth improves the numerical convergence of the third method by starting with a preliminary study of the poles of the matrix $\mathbf{X}^{-1}$. No doubt, in the future, the information gleaned from these studies, and perhaps others, will lead to a highly automated and reliable technique for full potential MST calculations.

4.7 The Green's function for an impurity embedded in a periodic lattice

There are interesting physics studies that can be modeled as a single impurity embedded in an otherwise perfect crystal. The impurity typically modifies the atoms on two or three nearest-neighbor shells, but we will start the discussion by assuming that the only effect is to modify the scattering matrix on the central site. The matrix $\mathbf{M}$, defined in equation (4.71), can be written

$$\mathbf{M} = \mathbf{M}^0 + \mathbf{d}, \tag{4.90}$$

where $\mathbf{M}^0$ is the matrix that describes the perfect crystal and has the same scattering matrix $\mathbf{m}^0$ on each site. The dimension of the perturbing matrix $\mathbf{d}$ is n, and it is written

$$\mathbf{d} = \mathbf{m}^1 - \mathbf{m}^0. \tag{4.91}$$

The differences in the free-electron propagators $(\mathbf{g}^{ij} - \mathbf{g}_0^{ij})$ will be ignored. In this model, the only change in the periodic lattice with $\mathbf{m}^0$ on every site is that that matrix is replaced with $\mathbf{m}^1$ on the central site.

Using standard matrix identities, the inverse of $\mathbf{M}$ is

$$\mathbf{M}^{-1} = (\mathbf{M}^0)^{-1} - (\mathbf{M}^0)^{-1}\mathbf{d}\mathbf{M}^{-1}. \tag{4.92}$$

It is easy to project out the scattering path matrix for the central site of the crystal,

$$\boldsymbol{\tau}^{11} = (\mathbf{I} + \boldsymbol{\tau}_0^{11}(\mathbf{m}^1 - \mathbf{m}_0))^{-1}\boldsymbol{\tau}_0^{11} = \boldsymbol{\tau}_0^{11}(\mathbf{I} + (\mathbf{m}^1 - \mathbf{m}_0)\boldsymbol{\tau}_0^{11})^{-1}. \tag{4.93}$$

The Green's function for the crystal with one impurity is now

$$G_1(E, \mathbf{r}, \mathbf{r}') = \mathbf{Z}^1(E, \mathbf{r}) \cdot \boldsymbol{\tau}^{11} \cdot \tilde{\mathbf{Z}}^{1\bullet}(E, \mathbf{r}') - \mathbf{Z}^1(E, \mathbf{r}) \cdot \tilde{\mathbf{J}}^{1\bullet}(E, \mathbf{r}'), \tag{4.94}$$

where the wave functions inside and outside the volume Ω_1 are obtained using the potential function $v_1(\mathbf{r})$ that leads to the scattering matrix $\mathbf{m}$. The scattering path operator for the host crystal is

$$\boldsymbol{\tau}_0^{11}(E) = \frac{\Omega}{(2\pi)^3}\int [\mathbf{m}_0(E) - \mathbf{g}(E, \mathbf{k})]^{-1} d\mathbf{k}, \tag{4.95}$$

which we know how to calculate.

It is not difficult to expand this theory to include changes in the scattering matrices on more than one site and also the change in propagation matrices due to displacements of the atoms from the sites dictated by the host crystal. The matrix we have called $\mathbf{d}$ simply becomes larger. However, the numerical calculations become more complex very rapidly.

References

[1] Wang Y (unpublished notes)
[2] Rusanu A, Stocks G M, Wang Y and Faulkner J S 2011 *Phys. Rev.* B **84** 035102
[3] Liu X, Wang Y, Eisenbach M and Stocks G M 2018 *Computer Physics Communications* **224** 265

Chapter 5

MST for systems with no long range order

To this point, all of the formulae we have derived for dealing with large collections of atoms assume that they are in periodic arrays. Bloch's theorem holds for such crystalline systems that have long range order, which means that attention can be focused on just one unit cell. The wave functions or Green's functions satisfy Bloch boundary conditions on the surface of the cell. This is obviously much easier than having to deal with all of the atoms in the system individually.

Aspects of condensed matter that are of interest in materials science include such phenomena as alloy phase diagrams, mechanical properties, growth processes and engineered materials. Understanding the mechanical properties of structural alloys requires the ability to do calculations on hundreds or thousands of atoms so that grain boundaries and dislocation cores can be simulated. In many of the most useful alloys, such as steels, the atoms have magnetic moments. Therefore, the effects of noncollinear magnetism must be understood. All of these studies require the ability to treat large systems of atoms with no long range order.

5.1 The supercell method

The band theory equation in chapter 3 is written for the case of one atom per unit cell. Equation (3.58) can easily be extended to handle any number of atoms per unit cell by generalizing the Korringa matrix to,

$$\mathbf{M}^{ij}(E, \mathbf{k}_n) = -\alpha \mathbf{c}^i(E)\mathbf{s}^j(E)^{-1}\delta_{ij} - \mathbf{B}^{ij}(E, \mathbf{k}_n), \tag{5.1}$$

where the superscripts refer to the different atoms in the cell. A large unit cell can be used to model any structure that we might want to model. We place the atoms wherever we want within a cube, as sketched in figure 5.1.

These cubes are then reproduced to fill all space. Using the above equation to treat the resulting periodic structure has the same effect as using periodic boundary conditions on the cube.

doi:10.1088/2053-2563/aae7d8ch5

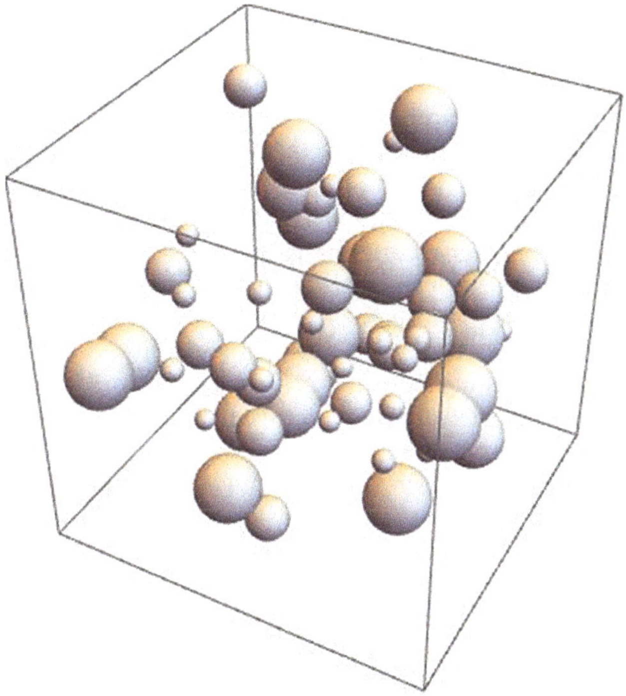

Figure 5.1. A drawing of three kinds of atoms placed randomly within a cube.

This method for modeling a structure is known as the supercell method. The difficulty with this method is that experience has shown that the time required for a calculation increases like the cube of the number N of the atoms in the unit cell, $T \propto N^3$. No matter how many computer resources are available, this exponential increase will eventually make the calculation impractical.

5.2 An order-N method for large systems

While theoretically and computationally non-trivial, the development of a truly order-N method represents a significant advance in quantum mechanical simulation methods. An order-N method is one for which the computational work grows linearly with system size, $T \propto N$, rather than exponentially. The locally self-consistent multiple scattering (LSMS) method is such an order-N approach [1]. It is based on the observation that a good approximation to the electron density and the density of states and charge density on a particular atom within a large system, and thereby the total energy of that system, can be obtained by considering only the electronic multiple scattering processes in a finite spatial region centered at that atom.

To implement the LSMS method, we first construct a picture of the structure we want to emulate and place the atoms on the proper sites. We developed the method for calculating the Green's function for a large number of atoms in chapter 4. The problem with calculating the **M** matrix as in equation (4.71) is that it is infinite dimensional. In the LSMS, we focus on one atom at a time, say the nth atom, and draw a sphere around it. The atoms inside this sphere are said to be within the local interaction zone (LIZ), as illustrated in figure 5.2.

We make a first guess at the potentials for the atoms in the LIZ, and calculate the scattering matrices $m^{i}_{LL'}(E)$ for each of them. We also calculate the free-particle Green's functions $g^{ij}_{LL'}(E, \mathbf{R}_{ij})$ that connect them. We calculate a matrix $\mathbf{M}^{n}$, which is like the matrix in equation (4.71) except that it has been made finite by ignoring the atoms outside of the LIZ. The scattering path matrix, $\boldsymbol{\tau}^{nn}(E)$, is the n,n element of the inverse of $\mathbf{M}^{n}$. Next, we calculate the Green's function as in equation (4.64). This process must be repeated for a range of energies and for every atom in the model. From that information, we can obtain the charge density, the next iteration to the potential, and the Madelung potential for each atom in the model. The entire calculation must be repeated until self-consistency is obtained, but it is still not completed. It would be best to find the forces on the atoms, because our original positioning of the atoms was a first guess. It might be necessary to rearrange them into equilibrium positions. The entire calculation will then have to be repeated.

The process outlined in the preceding paragraph sounds like a great deal of work. Many of the steps would be the same if, for example, the supercell method was used. As complicated as it is, the LSMS retains the advantage that it is order-N. Starting with the original publication of the method in 1995, many numerical tests have been carried out which demonstrate this fact. One is shown in figure 5.3.

The fundamental assumption that excellent approximations to the charge density and DOS on a given atom can be obtained from a calculation that ignores atoms

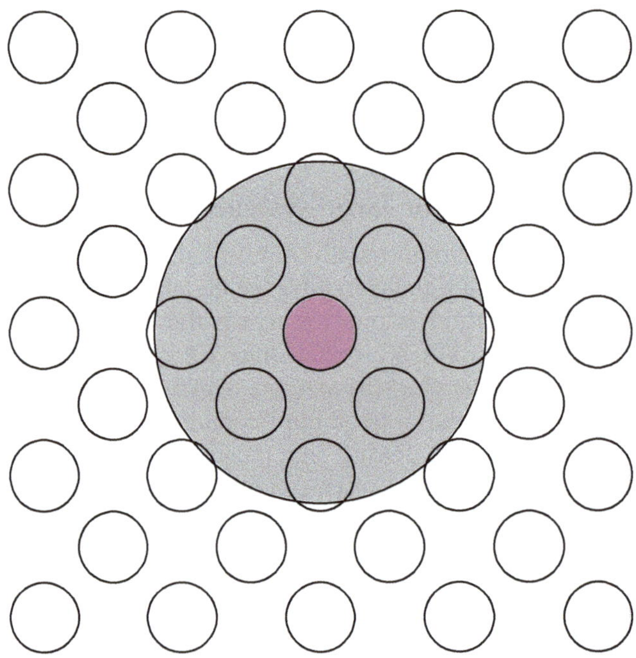

Figure 5.2. A sketch illustrating the central atom and the local interaction zone (gray).

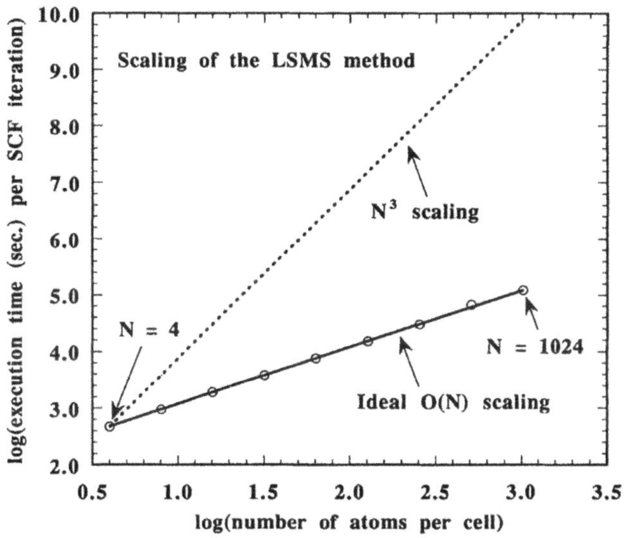

Figure 5.3. Scaling behavior of the LSMS algorithm. The calculation is for a large cell model of fcc Cu. The data were gathered for a LIZ consisting of 13 atoms, sufficient for determining the SCF charge densities. The open circles are the logarithm of the execution time per SCF iteration versus the logarithm of the system size. The solid line shows an O(N) scaling, and the dotted line shows an O(N^3) scaling.

outside the LIZ has also been tested. In practice, when applying the LSMS method to an *ab initio* electronic structure calculation for a system containing more than one kind of atom, one has the freedom to allow the LIZ for each atom to have a different size. A major question that is frequently asked is: what is the proper choice of the LIZ size for an atomic site? A related question is: as the LIZ size increases, will the LSMS results converge to the exact results?

Band theory can be used to calculate the properties of an atom in a periodic metallic solid to great accuracy. The total energy of an atom in FCC Cu and BCC Mo versus the LIZ radius have been calculated using the full potential LSMS method. For FCC Cu, the LSMS energy agrees with the band theory calculation to better than 0.5 mRyd when 6 neighboring shells are included in the LIZ. This corresponds to a cluster of 87 atomic sites with a LIZ radius of 11.7 a.u. For BCC Mo, on the other hand, a larger LIZ is required in order to achieve better than 0.5 mRyd accuracy. This is due to the fact that the Fermi energy falls in the d bands so that the density of states near the Fermi energy is significant. It is found that including 10 neighboring shells, corresponding to 137 atoms or a radius of 14.2 a.u., is necessary.

The question as to whether the LSMS is a practical way to study a problem in condensed matter physics or materials science depends on the computing environment that is available. Clearly, the LSMS is highly scalable on a massively parallel processing (MPP) supercomputer since each computer node can be assigned the calculation of the scattering matrix elements, the electron density, and the density of states for the atoms mapped onto it. A schematic representation of the mapping of the LSMS algorithm onto a MPP platform is displayed in figure 5.4, where, in the interests of clarity, we assume a one atom per node implementation scheme.

MPP computers are becoming more widely available. They now have computing power in the tens of petaflops with hundreds of thousands of nodes. With the addition of graphics processing units (GPUs) to more highly evolved central processing units (CPUs) it is possible to achieve speeds of hundreds of petaflops, and heading toward the goal of exaflops.

The LSMS method has been implemented for system sizes containing tens of thousands of atoms, limited only by the number of processors in the MPP platform. As the machines become faster, it is practical to assign more than one atom to a node. Since 1995, the LSMS has been used to elucidate many questions in physics and materials science that involve complex structures with atomic and positional disorder.

A question that can be answered almost trivially by LSMS calculations had never been understood before. This is the Coulomb energy in alloys. Experimental studies on core level chemical shifts in alloys were easily and accurately explained for the first time [2]. A relationship between the charge on an atomic site in an alloy and the Madelung potential at that site was discovered, [3] as shown in figure 5.5. Arguments can be given for the proportionality between the charge and the potential, but the extreme accuracy with which the straight line is tracked is surprising.

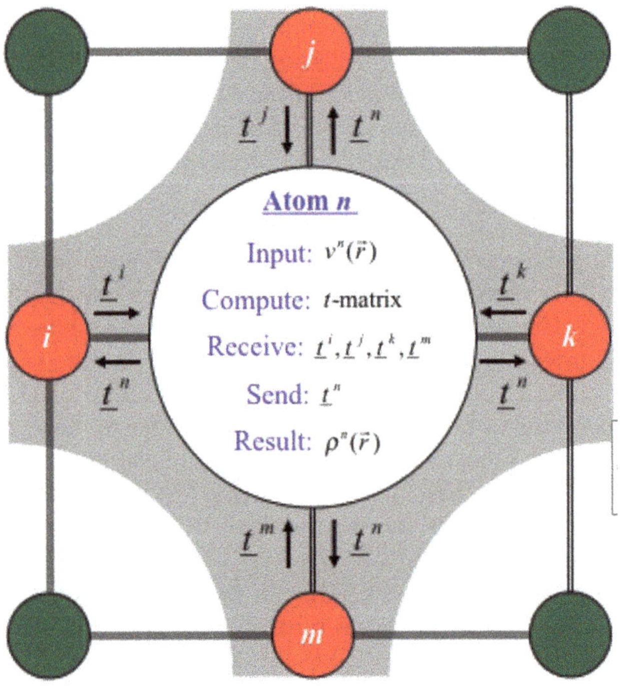

Figure 5.4. The communication pattern of the LSMS method implemented on parallel processors with one-atom to one-processor mapping. The gray area depicts the LIZ region of atom *n* which includes atoms *i, j, k, m*, and *n* itself. The same computation and communication activities take place simultaneously on all the other processors.

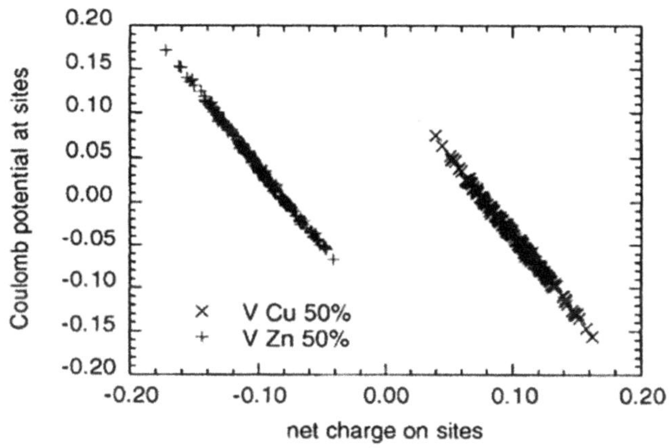

Figure 5.5. The Coulomb potential V' versus the charges q' on the 432 sites in the cell for a 50% Cu–Zn alloy with a bcc crystal structure. The plus signs are the data points for sites that have a Zn atom on them and the crosses are for Cu. The potential and charges are in dimensionless atomic units. Reprinted with permission from [3]. ©1995 American Physical Society.

5.3 Magnetism

Magnetism in solids is caused by the exchange interaction, which is a consequence of the Fermi statistics that governs the state of electrons. In ionic compounds that contain certain transition metals or rare earths, this leads to a picture of localized spins interacting by an exchange coupling. It is called a Heisenberg model, and Monte Carlo calculations on this model produce a Curie–Weiss law for the magnetic susceptibility that terminates in a transition temperature T_c at which the spins become ordered.

It is more difficult to develop a theory of metals that have itinerant electrons with energies in bands. Stoner, Wohlfarth, Slater, and others developed a band theory of magnetism in which there is a spin up and a spin down band that are displaced in energy relative to each other. The common Fermi energy means that there will be a net spin magnetism in the solid. The modern form of this theory relies on a spin-density version of the LDA, and predicts accurately that some metals will be magnetic and others will not. It gives a reasonably accurate prediction for the size of the magnetic moment on the atoms. It fails totally in predicting the behavior to magnetic metals at finite temperature. The only way that magnetism changes with temperature in this model, and disappears at some critical temperature T_c, is due to the thermal smearing of the Fermi distribution function. This leads to a physical picture that the spins in a magnetic metal are either all aligned or they cease to exist, a result that is contradicted by neutron diffraction experiments. It also leads to unrealistically high values for T_c.

The challenge in real metallic systems is to develop a theory that combines elements of the Stoner model with those of a Heisenberg model. This goal was achieved theoretically by the Hubbard model. Solutions of this simple model have the desired features, but the model cannot be checked experimentally because it contains only heuristic adjustable parameters. In 1985, Gyorffy *et al* [4] began a program for combining the Hubbard model with multiple scattering theory to achieve an *ab initio* theory of magnetism in metals. This theory is based on a

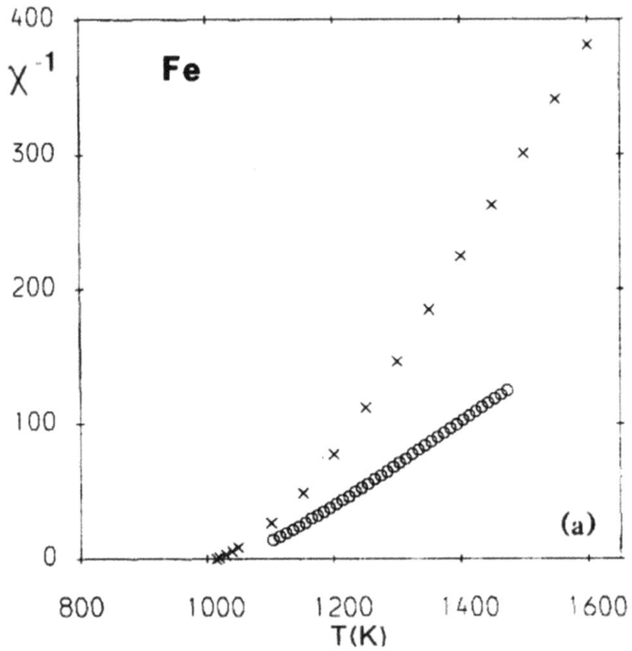

Figure 5.6. The temperature dependence of the inverse paramagnetic susceptibility of bcc iron in atomic units. The crosses show the theoretical results and the circles show the experimental data. Reprinted with permission from [5]. ©1992 American Physical Society.

treatment of finite temperature fluctuations in the moments and exchange fields using the modern techniques of statistical mechanics.[1] It has been called the mean-field disordered local moment (DLM) model. It represents a huge advance in the description of finite temperature magnetism in metals, but the reliance on mean-field theory at some parts of the derivation put a limit on the accuracy that can be achieved. A method to improve the mean-field approximation that replaces the simple averaged environment of a spin with the Onsager cavity field has been used in other applications of statistical mechanics.

When the Onsager cavity field was used in connection with the *ab initio* DLM model [5], the results shown in figure 5.6 were obtained for the temperature dependence of the susceptibility of iron.

The theoretical value of T_c for iron, 1015 K, is in excellent agreement with the experiment, 1040 K. The susceptibility curves are not identical, but recall that there is no susceptibility curve for the Stoner theory.

Similar results for nickel are shown in figure 5.7. As usual, it is more difficult to treat the smaller and more elusive moments of nickel than those of iron. This is the best *ab initio* calculation that has been done, but the theoretical and experimental values of T_c, 450 K and 650 K, are not very close. However, the Stoner prediction of 3000 K is ridiculous.

[1] A similar approach was used by the same authors to predict concentration waves in metals, and is described in chapter 9.

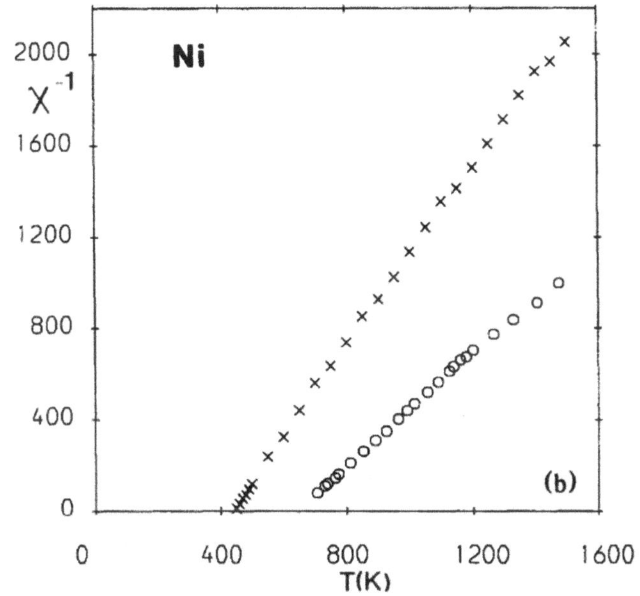

Figure 5.7. The temperature dependence of the inverse paramagnetic susceptibility of fcc nickel in atomic units. The crosses show the theoretical results and the circles show the experimental data. Reprinted with permission from [5]. ©1992 American Physical Society.

The DLM calculations shown above are far and away the best first principles calculations of magnetism in transition metals. They give a theory for moments in metals, but they do not give a picture of them. That had to wait for the development of the LSMS.

Noncollinear magnetic crystals are such that the magnetic moments associated with individual sites in the crystal are not aligned along the same axis of global magnetization. The LSMS can be extended to treat noncollinear magnetic systems. Calculations of the magnetic structure of a substitutionally disordered fcc $Fe_{65}Ni_{35}$ alloy indicate the existence of a noncollinear ground state. This calculation demonstrates the capability of the LSMS for treating magnetism, and it also clarifies questions about the alloy, called INVAR, that have been debated for many years [6].

Assuming that the exchange field orientation is sufficiently constant over some volume surrounding each atomic site, it is possible to define for each atom in the alloy a local, site dependent, frame of reference in which the local magnetic field aligns in the local 'up' or 'down' direction. In this local frame, standard methods for calculating the scattering matrix for a single scatterer can be applied. Since each site has its own local frame, unitary transformations are necessary in order to construct the scattering path matrix required when calculating the Green's function

$$\begin{aligned} G(E, \mathbf{r}, \mathbf{r}') = &\sum_{\sigma,\sigma'} \mathbf{Z}^{n\sigma}(E, \mathbf{r}) \cdot \boldsymbol{\tau}^{n\sigma n\sigma'}(E) \cdot \tilde{\mathbf{Z}}^{n\sigma' \bullet}(E, \mathbf{r}') \\ &- \sum_{\sigma} \mathbf{Z}^{n\sigma}(E, \mathbf{r}) \cdot \tilde{\mathbf{J}}^{n\sigma\bullet}(E, \mathbf{r}'). \end{aligned} \tag{5.2}$$

The Green's function differs from the ones used previously by the introduction of the spin coordinate, σ, which takes two values corresponding to up and down. The boldface variables are used to eliminate showing the sums over the angular momentum L. We know how to obtain the t-matrices and wave functions on each site because they are calculated as in the Stoner model. They are simply different for each site in the scattering path operator depending on the orientation of the local magnetic field.

Figure 5.8. shows the initial random magnetic moment orientations in a 256-atom model of a fcc Fe0.65Ni0.35 INVAR alloy. The random arrangement of moments corresponds to a very high temperature.

After the LSMS codes have been run to self-consistency, the spins are as shown in figure 5.9. A complex magnetic structure is found involving noncollinear configurations associated with Fe-rich regions. Most magnetic moments align ferromagnetically, i.e. toward the magnetization direction. A few sites, however, remain in which the magnetic moment is not parallel to the overall magnetization direction. One site is antiparallel and other sites have their magnetization canted with respect to the sample magnetization. Careful examination of these sites shows that these atoms have less than three Ni atoms in the first neighboring shell. For the single Fe atom in our sample that is surrounded completely by Fe atoms, the magnetic moment aligns antiferromagnetically. That is, its moment orientation is opposite to the magnetization direction. It is known that pure fcc iron is antiferromagnetic. For

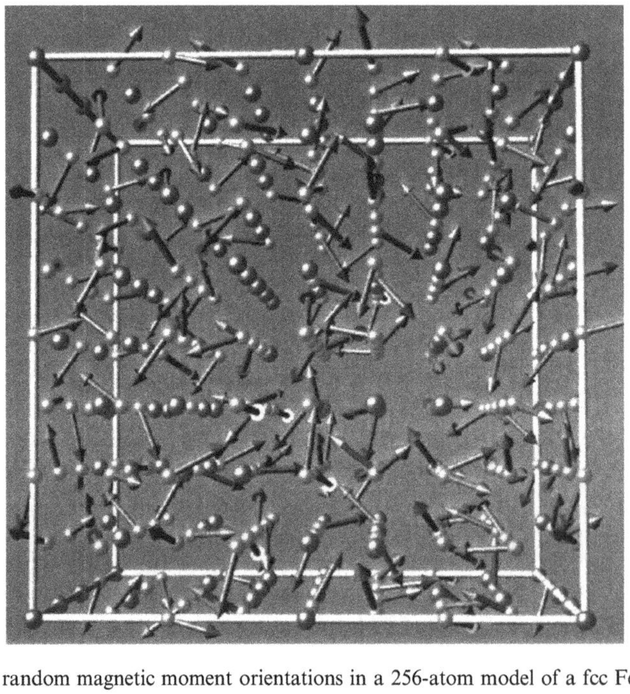

Figure 5.8. Initial random magnetic moment orientations in a 256-atom model of a fcc $Fe_{0.65}Ni_{0.35}$ INVAR alloy. The small spheres are Fe sites, and the large ones represent Ni. The magnitude and direction of the magnetic moments are indicated by the length and direction of the arrows.

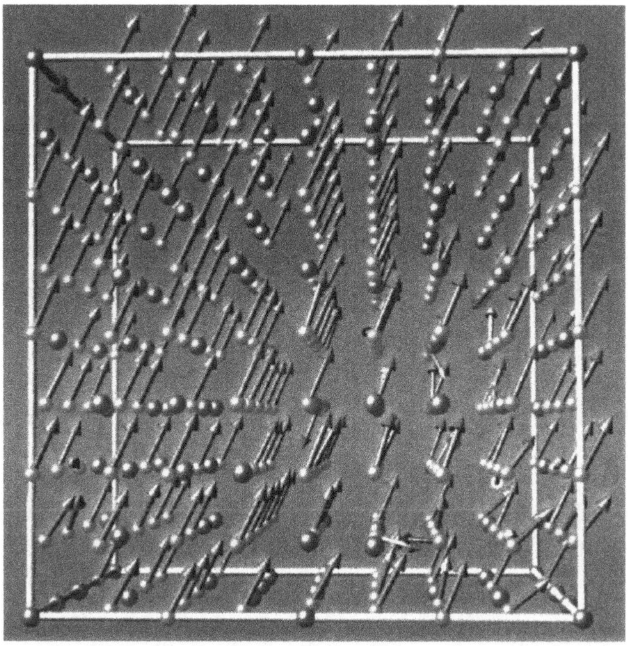

Figure 5.9. Final self-consistent magnetic moment orientations in a 256-atom model of a fcc Fe0.65Ni0.35 INVAR alloy. The magnitude and direction of the magnetic moments are indicated by the length and direction of the arrows.

Fe atoms having only one or two Ni atoms in their first neighboring shell, a variety of noncollinear magnetic moment arrangements are found. These results agree with the experiment.

Stocks *et al* [7] showed that the theory of spin dynamics theory of magnetism put forward by Antropov *et al* [8] contains a fundamental flaw. It overlooks the fact that an arbitrary noncollinear spin state, such as may be encountered in the simulation of the paramagnetic state, does not correspond to a LSDA extremum and hence is not well defined. This difficulty is overcome by restating the argument within the framework of the constrained local moment and the LSMS.

Other applications of the LSMS will be discussed in chapter 9.

5.4 The coherent potential approximation for random alloys

The LSMS method can be used to simulate any ordered or disordered system, including those with random atomic positions. Prior to that development, when computer power had not reached a point that LSMS calculations for large clusters of atoms could be contemplated, theories of electronic structure in systems without long range order focused on substitutional solid solution alloys. Such alloys are made up of two or more species of atoms distributed randomly on the sites $\mathbf{R}_i$ of a Bravais lattice. In the following, we concentrate on binary alloys, although the generalization to ternary, quaternary, etc, is straightforward.

The ensemble of all alloys that can be constructed out of $N_A = c_A N$ A atoms and $N_B = c_B N$ B atoms contains $M_c = N!/N_A!N_B!$ atomic arrangements. Experimentally, it is observed that all alloys with the concentrations c_A and c_B have approximately the same electronic structure. It follows that, as $N \to \infty$, there is an electronic structure that all alloys with the same concentrations will have with probability one, and it can be found by averaging over the ensemble. The ensemble average of a one-electron wave function has no meaning, but, because of the linear relation between the Green's function and the electronic properties, the ensemble average of the Green's function is useful. The formulation presented here closely follows a paper by Faulkner and Stocks [9], and that paper should be consulted for additional details.

The ensemble average of the function in equation (4.64) can be obtained if we assume that the resulting alloy model is isomorphous and we invoke the single site approximation. Isomorphism means that the Green's function takes one of two forms, $G_A(E, \mathbf{r}_n, \mathbf{r}'_n)$ if there is an A atom on site n, and $G_B(E, \mathbf{r}_n, \mathbf{r}'_n)$ if there is a B atom on that site. The single site, or mean field, approximation is more complicated. As a practical matter, it is equivalent to the CPA condition in equation (5.11). With these assumptions, the ensemble averaged Green's function can be written

$$\langle G_n(E, \mathbf{r}_n, \mathbf{r}'_n)\rangle = c_A G_A(E, \mathbf{r}_n, \mathbf{r}'_n) + c_B G_B(E, \mathbf{r}_n, \mathbf{r}'_n), \tag{5.3}$$

where

$$G_A(E, \mathbf{r}_n, \mathbf{r}'_n) = \mathbf{Z}^A(E, \mathbf{r}_n) \cdot \langle \boldsymbol{\tau}^{nn}\rangle_A(E) \cdot \tilde{\mathbf{Z}}^{A\bullet}(E, \mathbf{r}'_n) - \mathbf{Z}^A(E, \mathbf{r}_n) \cdot \tilde{\mathbf{J}}^{A\bullet}(E, \mathbf{r}'_n), \tag{5.4}$$

and

$$G_B(E, \mathbf{r}_n, \mathbf{r}'_n) = \mathbf{Z}^B(E, \mathbf{r}_n) \cdot \langle \boldsymbol{\tau}^{nn}\rangle_B(E) \cdot \tilde{\mathbf{Z}}^{B\bullet}(E, \mathbf{r}'_n) - \mathbf{Z}^B(E, \mathbf{r}_n) \cdot \tilde{\mathbf{J}}^{B\bullet}(E, \mathbf{r}'_n). \tag{5.5}$$

The implicit assumption in this equation is that $\mathbf{r}_n$ and $\mathbf{r}'_n$ are both within the nth Wigner–Seitz sell, Ω_n, and the matrix notation is explained in equations (4.41). The charge density for an A atom, $\rho_A(\mathbf{r}_n)$, is obtained by inserting $G_A(E, \mathbf{r}_n, \mathbf{r}_n)$ into equation (4.4), and similarly for the B atom. From these charge densities we calculate the potentials $v_A(\mathbf{r}_n)$ and $v_B(\mathbf{r}_n)$. The functions $Z_L^A(E, \mathbf{r}_n)$ and $J_L^A(E, \mathbf{r}_n)$ are obtained using $v_A(\mathbf{r}_n)$ in equation (4.34), and the functions $Z_L^B(E, \mathbf{r}_n)$ and $J_L^B(E, \mathbf{r}_n)$ are obtained using $v_B(\mathbf{r}_n)$.

The averages $\langle \boldsymbol{\tau}^{nn}\rangle_A$ and $\langle \boldsymbol{\tau}^{nn}\rangle_B$ are the ensemble averages of the scattering path operator subject to the condition that there is an A or B atom on site n. These averages are simplified considerably by making the single site approximation. It is assumed that, on average, the scattering of the electrons from all the sites other than site n is described by replacing the actual potentials on those sites with a potential $v_c(\mathbf{r})$, called the coherent potential. The scattering from $v_c(\mathbf{r})$ is described by a t-matrix, $\mathbf{t}_c(E)$, which has an inverse $\mathbf{m}_c(E)$. It follows that the ensemble averages are the matrices obtained from equation (4.93)

$$\langle \boldsymbol{\tau}^{nn}\rangle_A = [\mathbf{I} + \boldsymbol{\tau}_c^{nn}(\mathbf{m}_A - \mathbf{m}_c)]^{-1}\boldsymbol{\tau}_c^{nn}, \tag{5.6}$$

and

$$\langle \boldsymbol{\tau}^{nn} \rangle_B = \left[\mathbf{I} + \boldsymbol{\tau}_c^{nn} (\mathbf{m}_B - \mathbf{m}_c) \right]^{-1} \boldsymbol{\tau}_c^{nn}. \tag{5.7}$$

The matrices $\mathbf{m}_A$ and $\mathbf{m}_B$ are the inverses of the t-matrices calculated from the potentials $v_A(\mathbf{r})$ and $v_B(\mathbf{r})$. The matrix $\boldsymbol{\tau}_c^{nn}$ describes a periodic system, so it takes the form of equation (4.74)

$$\boldsymbol{\tau}_c^{nn}(E) = \frac{\Omega}{(2\pi)^3} \int [\mathbf{m}_c(E) - \mathbf{g}(E, \mathbf{k})]^{-1} d\mathbf{k}. \tag{5.8}$$

The matrices in these equations have dimension $(l_{\max} + 1)^2$. None of the equations actually depend on the choice of the site n.

It will be useful later to note that equations (5.6) and (5.7) may be written

$$\langle \boldsymbol{\tau}^{nn} \rangle_X = \mathbf{D}_X \boldsymbol{\tau}_c^{nn} = \boldsymbol{\tau}_c^{nn} \tilde{\mathbf{D}}_X, \tag{5.9}$$

with

$$\mathbf{D}_X = \left[\mathbf{I} + \boldsymbol{\tau}_c^{00} (\mathbf{m}_X - \mathbf{m}_c) \right]^{-1} \qquad \tilde{\mathbf{D}}_X = \left[\mathbf{I} + (\mathbf{m}_X - \mathbf{m}_c) \boldsymbol{\tau}_c^{00} \right]^{-1}. \tag{5.10}$$

In the language of multiple scattering theory used here, the CPA condition put forward by Soven [19] that defines the t-matrix $\mathbf{t}_c(E)$ is

$$c_A \langle \boldsymbol{\tau}^{nn} \rangle_A + c_B \langle \boldsymbol{\tau}^{nn} \rangle_B = \boldsymbol{\tau}_c^{nn}. \tag{5.11}$$

In other words, this is the condition where, on average, there is no scattering from A and B atoms embedded in the coherent potential host matrix. It should be emphasized at this point that, although we speak of a coherent potential $v_c(\mathbf{r})$, that object never appears explicitly in the theory. We certainly never put $v_c(\mathbf{r})$ in a Schrödinger equation and solve for wave functions. The potentials $v_A(\mathbf{r}_n)$ and $v_B(\mathbf{r}_n)$ depend on each other and on the concentrations. This information is transferred through the scattering matrix $\mathbf{t}_c(E)$. Referring to equation (5.3), we see that the condition in equation (2.11) is equivalent to the statement that the ensemble average Green's function can be treated as the Green's function for a hypothetical lattice of CPA scatterers

$$\langle G(E, \mathbf{r}, \mathbf{r}') \rangle = G_c(E, \mathbf{r}, \mathbf{r}'). \tag{5.12}$$

Space does not allow for descriptions of all the theoretical arguments put forward to justify the CPA condition. We will give a few of them in chapter 7 in connection with model calculations. The best corroboration for the method is that it has been used with great success to explain a number of phenomena in condensed matter physics and materials science.

The procedure for calculating the electronic states within the CPA is first to guess the starting values for the charge densities $\rho_A(\mathbf{r})$ and $\rho_B(\mathbf{r})$ and from them calculate $v_A(\mathbf{r})$ and $v_B(\mathbf{r})$. The t-matrices $\mathbf{t}_A(E)$ and $\mathbf{t}_B(E)$ are calculated, and a first guess to $\mathbf{t}_c(E)$ is just the average of them $\mathbf{t}_c(E) \approx c_A \mathbf{t}_A(E) + c_B \mathbf{t}_B(E)$. Equations (5.6)–(5.11) are then iterated to self-consistency in the charge densities and also in the coherent potential t-matrix $\mathbf{t}_c(E)$.

Using the standard formula for calculating the density of states function from the Green's function in equation (4.5), we obtain from equation (5.3) that the DOS for per unit cell in the alloy is

$$\rho_c(E) = c_A\rho_A(E) + c_B\rho_B(E), \tag{5.13}$$

with

$$\rho_A(E) = -\frac{2}{\pi}\mathrm{Im}\int_{\text{unit cell}} G_A(E, \mathbf{r}, \mathbf{r})d\mathbf{r}, \quad \rho_B(E) = -\frac{2}{\pi}\mathrm{Im}\int_{\substack{\text{unit}\\ \text{cell}}} G_B(E, \mathbf{r}, \mathbf{r})d\mathbf{r}. \tag{5.14}$$

It is useful to know how the DOS is partitioned between that which is associated with A atoms and B atoms. Interestingly, that is the natural way to calculate it. Once the DOS is known, the IDOS and hence the Fermi energy E_F can be found.

In addition to the formulas for densities of states based on Green's functions, an equation based on a formula put forward by Lloyd [10] can be used to calculate the integrated density of states

$$\begin{aligned} N(E) = N_0(E) &- \frac{1}{\pi}\,\mathrm{Im}\,\frac{1}{\Omega_{BZ}}\left[\int_{BZ} \ln\det \mathbf{M}_c(E, \mathbf{k})d\mathbf{k}\right. \\ &\left. - c_B \ln\det \frac{\mathbf{m}_A - \langle\mathbf{m}\rangle}{\mathbf{m}_A - \mathbf{m}_c} - c_A \ln\det \frac{\mathbf{m}_B - \langle\mathbf{m}\rangle}{\mathbf{m}_B - \mathbf{m}_c}\right] \end{aligned} \tag{5.15}$$

where

$$\mathbf{M}_c(E, \mathbf{k}) = \mathbf{m}_c(E) - \mathbf{g}(E, \mathbf{k}). \tag{5.16}$$

Lloyd's formula will be discussed further in the following chapters. The fact that the numerical derivative of this formula leads to the same result as equation (5.13) lends credence to both formulas.

The reason that copper–nickel alloys were chosen for this early calculation is that the first success of the CPA in explaining a real experiment was for this system [11]. The calculated DOS in figure 5.10, particularly the split off band in the 77% copper alloy, provide the first theoretical explanation for the measurements of this quantity using ultraviolet photoemission spectroscopy. This success encouraged the use of the fledgling CPA theory.

All of the early calculations using the CPA were done on simplified models, such as the tight-binding approximation. The MST formalism for the CPA described above uses much the same technology as the KKR band theory method of chapter 3. Gyorffy and Stocks were the first to apply these equations without approximations [12]. Their Korringa–Kohn–Rostoker-Coherent-Potential-Approximation (KKR-CPA) is a set of equations and computer programs that is now the gold standard for CPA calculations. The computer programs continue to be improved and adapted for high performance computers.

Some of the more spectacular successes of the KKR-CPA are outlined below and in chapter 9. The resistivity of alloys has been accurately predicted [13]. The propensity of a random alloy to transform into a superlattice structure or into an

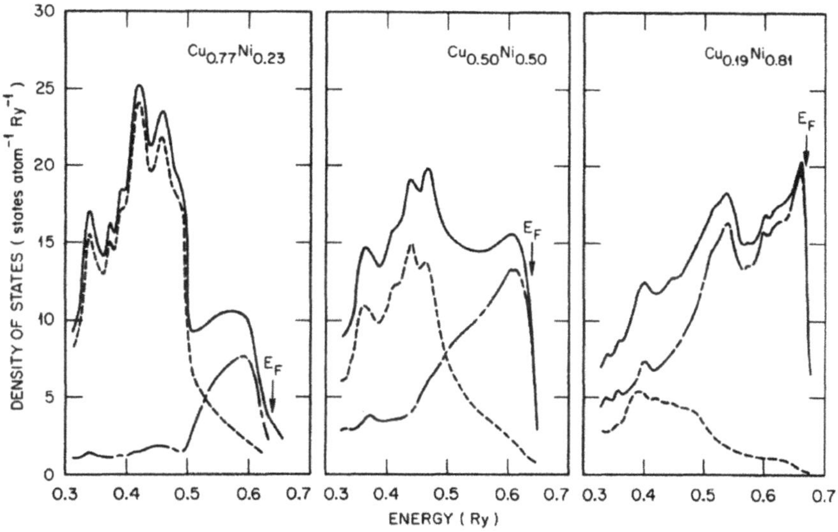

Figure 5.10. The partitioned densities of states and the total density of states for three copper–nickel alloys with the concentrations shown [12]. The total density of states is also calculated by numerically differentiating the formula for the integrated density of states in equation (5.15). Reprinted with permission from [12]. ©1980 American Physical Society.

incommensurately ordered structure has been successfully predicted [14]. The KKR-CPA has proved to be important in the development of theories of high temperature magnetism in metals and alloys [15].

5.5 The spectral density function

The equations in the preceding section that focus on a single site are sufficient for calculating the important CPA parameters like $\mathbf{t}_c(E)$, E_F, $\rho_A(\mathbf{r})$, and $\rho_B(\mathbf{r})$. Once that has been accomplished, there is an ancillary function called the spectral density that can be obtained. It is a very useful quantity that was first derived correctly in reference [9], where it is called the Bloch spectral density function, $A(E, \mathbf{k})$. With this function we reintroduce the Bloch vectors $\mathbf{k}$ and Brillouin zones familiar in band theory calculations. This can be done because the ensemble averaged Green's function describes a system that is invariant under the space group of the underlying Bravais lattice, although no single element of the ensemble has that property.

It is only a matter of algebra to show that a Green's function is diagonal in the k-representation when it is periodic in real space $G_c(E, \mathbf{r} + \mathbf{R}_n, \mathbf{r}' + \mathbf{R}_n) = G_c(E, \mathbf{r}, \mathbf{r}')$. The formula for the Bloch Fourier transform of $G_c(E, \mathbf{r}, \mathbf{r}')$ is

$$\hat{G}(E, \mathbf{k}) = \sum_n e^{i\mathbf{k}\cdot\mathbf{R}_n} \int_{\Omega_0} G_c(E, \mathbf{r}_0, \mathbf{r}_0 + \mathbf{R}_n) d\mathbf{r}_0. \tag{5.17}$$

In order to evaluate $G_c(E, \mathbf{r}_0, \mathbf{r}_0 + \mathbf{R}_n)$ with $\mathbf{R}_n \neq 0$ within the CPA, it is necessary to develop a non-site-diagonal (NSD) version of the ensemble averaged Green's function. The site-diagonal (SD) version is in equation (5.5). Lengthy arguments were put forward in reference [9] for the formula

$$
\begin{aligned}
G_c(E, \mathbf{r}_n, \mathbf{r}'_m) &= c_A^2 \mathbf{Z}^A(E, \mathbf{r}_n) \cdot \mathbf{D}_A \cdot \boldsymbol{\tau}_c^{nm} \cdot \tilde{\mathbf{D}}_A(E) \cdot \tilde{\mathbf{Z}}^{A\bullet}(E, \mathbf{r}'_m) \\
&+ c_A c_B \mathbf{Z}^A(E, \mathbf{r}_n) \cdot \mathbf{D}_A \cdot \boldsymbol{\tau}_c^{nm} \cdot \tilde{\mathbf{D}}_B(E) \cdot \tilde{\mathbf{Z}}^{B\bullet}(E, \mathbf{r}'_m) \\
&+ c_B c_A \mathbf{Z}^B(E, \mathbf{r}_n) \cdot \mathbf{D}_B \cdot \boldsymbol{\tau}_c^{nm} \cdot \tilde{\mathbf{D}}_A(E) \cdot \tilde{\mathbf{Z}}^{A\bullet}(E, \mathbf{r}'_m) \\
&+ c_B^2 \mathbf{Z}^B(E, \mathbf{r}_n) \cdot \mathbf{D}_B \cdot \boldsymbol{\tau}_c^{nm} \cdot \tilde{\mathbf{D}}_B(E) \cdot \tilde{\mathbf{Z}}^{B\bullet}(E, \mathbf{r}'_m),
\end{aligned}
\tag{5.18}
$$

where $\mathbf{D}_A$ and $\mathbf{D}_B$ are given by equation (5.10) and

$$
\boldsymbol{\tau}_c^{nm}(E) = \frac{\Omega}{(2\pi)^3} \int e^{i\mathbf{k}\cdot(\mathbf{R}_m - \mathbf{R}_n)} [\mathbf{m}_c(E) - \mathbf{g}(E, \mathbf{k})]^{-1} d\mathbf{k}. \tag{5.19}
$$

The function defined here has the symmetry of the reciprocal lattice

$$
\hat{G}(E, \mathbf{k} + \mathbf{K}_n) = \hat{G}(E, \mathbf{k}),
$$

where $\mathbf{K}_n$ is the translation vector that connects one Brillouin zone with another.

We assert that the spectral density function is

$$
A(E, \mathbf{k}) = -1/\pi \mathrm{Im} \hat{G}(E, \mathbf{k}). \tag{5.20}
$$

It gives the density of state vectors that have energy between E and $E + dE$ and k-vectors in the range $d\mathbf{k}$ around $\mathbf{k}$. The result of applying the same arguments to a periodic lattice leads to

$$
A(E, \mathbf{k}) = \sum_n \delta(E - E_n(\mathbf{k})), \tag{5.21}
$$

where the $E_n(\mathbf{k})$ are the energy eigenvalues found as the roots of the KKR matrix in equation (3.58). Since the peaks that define the constant energy surfaces in periodic systems are infinitely sharp, we expect $A(E, \mathbf{k})$ for a disordered alloy to have peaks but that they will be considerably less sharp.

The calculation of $A(E, \mathbf{k})$ can be made much easier by defining the row and column matrices

$$
\begin{aligned}
\mathbf{Z}^c(E, \mathbf{r}_n) &= c_A \mathbf{Z}^A(E, \mathbf{r}_n) \cdot \mathbf{D}_A + c_B \mathbf{Z}^B(E, \mathbf{r}_n) \cdot \mathbf{D}_B \quad \text{row matrix} \\
\tilde{\mathbf{Z}}^{c\bullet}(E, \mathbf{r}'_m) &= c_A \tilde{\mathbf{D}}_A(E) \cdot \tilde{\mathbf{Z}}^{A\bullet}(E, \mathbf{r}'_m) \\
&\quad + c_B \tilde{\mathbf{D}}_B(E) \cdot \tilde{\mathbf{Z}}^{B\bullet}(E, \mathbf{r}'_m) \quad \text{column matrix.}
\end{aligned}
\tag{5.22}
$$

Then the Green's function in equation (5.18) is identical to

$$
G_c(E, \mathbf{r}_n, \mathbf{r}'_m) = \mathrm{Tr}(\mathbf{F}^{cc}(E, \mathbf{r}_n, \mathbf{r}'_m) \cdot \boldsymbol{\tau}_c^{nm}), \tag{5.23}
$$

where Tr means the trace of and the square matrix is

$$
\mathbf{F}^{cc}(E, \mathbf{r}_n, \mathbf{r}'_m) = \tilde{\mathbf{Z}}^{c\bullet}(E, \mathbf{r}'_m) \cdot \mathbf{Z}^c(E, \mathbf{r}_n). \tag{5.24}
$$

Similarly, the SD Green's function in equation (5.3) is the same as

$$G_c(E, \mathbf{r}_n, \mathbf{r}'_n) = \mathrm{Tr}(\mathbf{F}^c(E, \mathbf{r}_n, \mathbf{r}'_n) \cdot \boldsymbol{\tau}_c^{nn}) \\ - \left(c_A \mathbf{Z}^A(E, \mathbf{r}_n) \cdot \tilde{\mathbf{J}}^{A\bullet}(E, \mathbf{r}'_n) + c_B \mathbf{Z}^B(E, \mathbf{r}_n) \cdot \tilde{\mathbf{J}}^{B\bullet}(E, \mathbf{r}'_n)\right), \tag{5.25}$$

using the square matrix

$$\mathbf{F}^c(E, \mathbf{r}_n, \mathbf{r}'_n) = c_A \tilde{\mathbf{Z}}^{A\bullet}(E, \mathbf{r}'_n) \cdot \mathbf{Z}^A(E, \mathbf{r}_n) \cdot \mathbf{D}_A \\ + c_B \tilde{\mathbf{Z}}^{B\bullet}(E, \mathbf{r}'_n) \cdot \mathbf{Z}^B(E, \mathbf{r}_n) \cdot \mathbf{D}_B. \tag{5.26}$$

It is necessary to deal with both the SD and NSD forms of the CPA Green's functions, because the sum in equation (5.17) includes the term with $\mathbf{R}_n = 0$.

Defining

$$\mathbf{F}^{cc}(E) = \int_{\Omega_0} \mathbf{F}^{cc}(E, \mathbf{r}_0, \mathbf{r}_0) d\mathbf{r}_0 \\ \mathbf{F}^c(E) = \int_{\Omega_0} \mathbf{F}^c(E, \mathbf{r}_0, \mathbf{r}_0) d\mathbf{r}_0, \tag{5.27}$$

and assuming the ZJ terms can be ignored because they are real, the final expression for $A(E, \mathbf{k})$ is

$$A(E, \mathbf{k}) = -1/\pi \mathrm{ImTr} \mathbf{F}^{cc} \boldsymbol{\tau}_c(E, \mathbf{k}) - 1/\pi \mathrm{ImTr}(\mathbf{F}^{cc} - \mathbf{F}^c) \boldsymbol{\tau}_c^{00}(E). \tag{5.28}$$

The complicated term to calculate is

$$\boldsymbol{\tau}_c(E, \mathbf{k}) = \sum_n e^{i\mathbf{k}\cdot\mathbf{R}_n} \boldsymbol{\tau}_c^{0n}, \tag{5.29}$$

but experience has shown that all of the terms are important.

5.6 Resistivity

The most compelling argument for the correctness of the NSD CPA Green's function in equation (5.18) and the spectral function in equation (5.28) is that they have been used to treat a number of phenomena in condensed matter physics and have been extremely successful in providing a theoretical justification for the experimental measurements. An early application of the theory is a study of the electrical conductivity and thermo power of silver–palladium alloys [16]. In the early days of careful detailed measurements of the conductivity of pure metal crystals it was surprising to find that it appeared to increase without bound as the temperature approached absolute zero. The explanation for this lies in the band theory for electrons that is explained in chapter 3, which shows that the electronic wave functions have the same extent throughout the crystal with no change in magnitude. The story would be expected to be very different for disordered alloys which are not invariant under a space group. Indeed, experiments show that such alloys have a finite low temperature conductivity. Little was known about the electronic structure of alloys at the time, and a formula was derived on the assumption that they had a

Fermi surface like a pure crystal but that the electron quasiparticles had a finite lifetime. It turns out that the formula so derived

$$\sigma = \frac{2e^2}{3(2\pi)^3} \int \frac{dS_k}{\hbar v_k} v_k^2 \tau_k. \tag{5.30}$$

can be used when the terms are interpreted in a modern way.

The authors of reference [14] carried out three self-consistent KKR-CPA calculations on silver–palladium alloys with 20%, 50%, and 80% silver content. They obtained the CPA scattering matrix $\mathbf{t}_c(E)$ and the scattering matrices $\mathbf{t}_A(E)$ and $\mathbf{t}_B(E)$. In this process, they necessarily found the Fermi energy E_F. They then did constant energy searches of $A(E_F, \mathbf{k})$ from equation (5.28). The values of $\mathbf{k}$ ranged along a set of rays that connected the center of the Brillouin zone (Γ) to the faces of the zone. As many as eight thousand directions were used in the irreducible 1/48th of the fcc Brillouin zone. A small fraction of these searches lying in the symmetry planes are shown in figure 5.11.

The conductivity for each of the alloys is calculated by fitting the well-defined peaks on each ray n to a Lorentzian function to get the peak maximum $E_n(\mathbf{k})$ and the width $\Gamma_n(\mathbf{k})$. Repeating this process for thousands of rays gives a highly detailed picture of the constant energy surface. The quasi-particle velocity that goes into equation (5.30) is obtained from the numerical derivatives $v_n(\mathbf{k}) = \nabla_{\mathbf{k}}(E_n(\mathbf{k}))$. The lifetime is related to the inverse of the peak-width by the uncertainty relation $\tau_n(\mathbf{k}) = \hbar/\Gamma_n(\mathbf{k})$. The smeared out structure around the X point makes no contribution to the conductivity according to the logic of our assumptions.

With this interpretation of the calculated $A(E_F, \mathbf{k})$, we obtain a comparison between theory and experiment for the resistivity of the silver–palladium alloy system. Calculations were done for $Ag_xPd_{(1-x)}$ for $x = 0.2$, 0.3, 0.4, 0.45, 0.5, 0.55, 0.6, 0.7, and 0.8. The resistivity is, of course, the inverse of the conductivity

$$\rho(x) = 1/\sigma(x). \tag{5.31}$$

The results of these calculations are shown in figure 5.12.

The excellent agreement with experiment obtained with a very logical application of the spectral density function $A(E_F, \mathbf{k})$ was very encouraging. As will be discussed in chapter 9, this function has been used to explain other phenomena, and it continues to be successful.

5.7 The polymorphous CPA

A problem that is endemic to the standard version of the CPA described in the preceding sections is that the Coulomb effects are not treated correctly. Integrating the charge densities $\rho_A(\mathbf{r})$ and $\rho_B(\mathbf{r})$ and subtracting the nuclear charges Z_A and Z_B leads to the net charges on the A and B sites, q_A and q_B. It would seem that there should be a Madelung contribution to the potentials $v_A(\mathbf{r})$ and $v_B(\mathbf{r})$ as in ordered alloys, but the only Madelung potentials that are commensurate with the assumption of an isomorphous alloy are zero. The isomorphous model of an alloy with the charge q_A on every A atom and q_B on every B atom is not realistic. Using the order-N

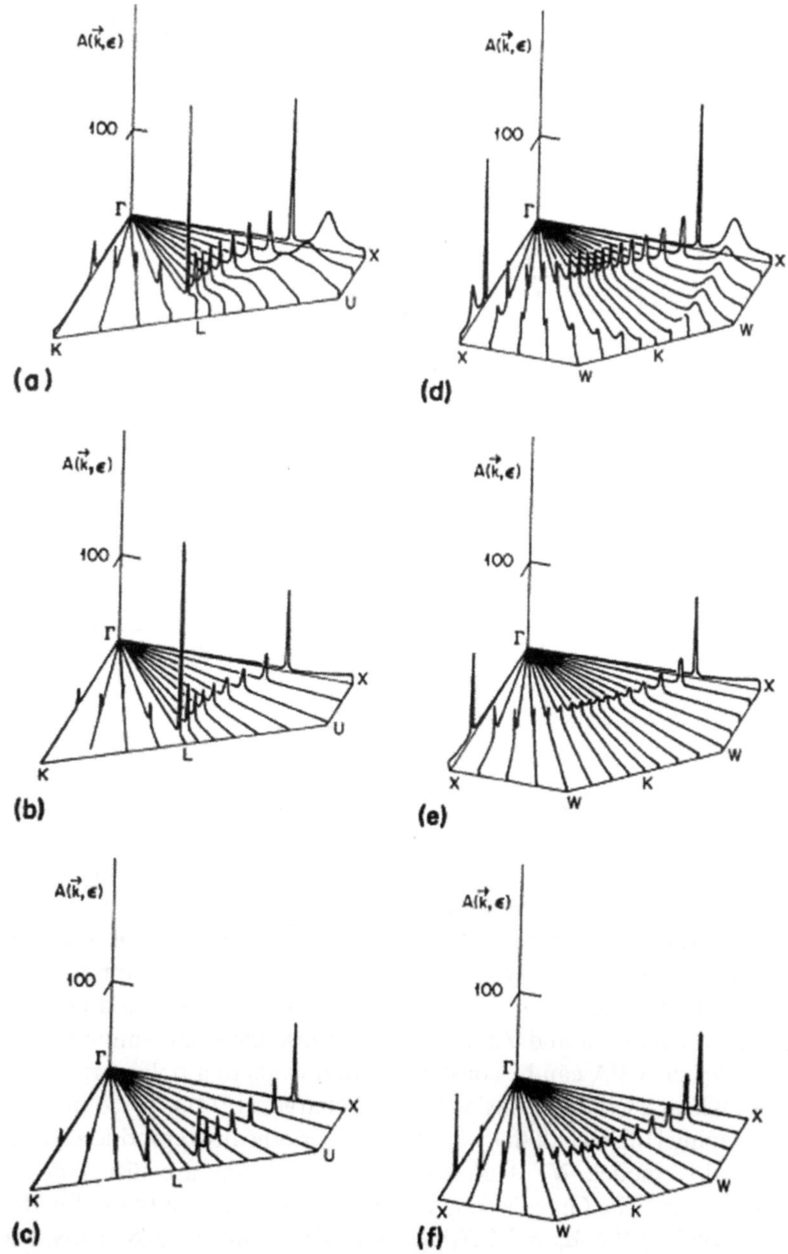

Figure 5.11. The results of constant energy calculations of the spectral density for silver–palladium alloys along directions lying on symmetry planes in an fcc Brillouin zone. For the 80% palladium alloy, panels (a) and (d), there is a well-defined constant energy surface centered around Γ, and a rather diffuse surface centered around X. For the 50% alloy, panels (b) and (e), the surface around Γ remains well defined. It has moved out toward the surface of the Brillouin zone. The surface centered around X has almost disappeared. In the silver-rich 20% palladium alloy, panels (c) and (f), the Γ centered surface is similar to the one in silver, touching the Brillouin zone at the point L.

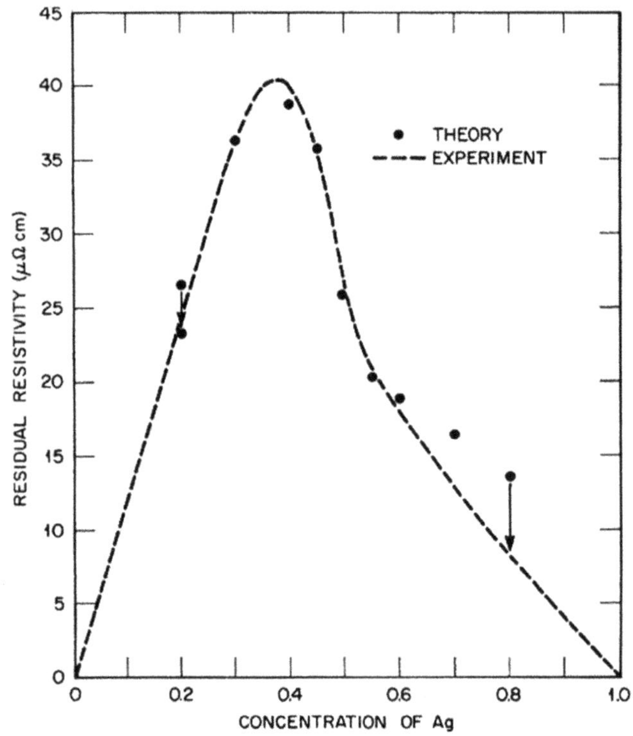

Figure 5.12. The residual resistivity of $Ag_xPd_{(1-x)}$ as a function of x. The arrow at $x = 0.8$ shows an estimate of the correction that would be obtained if a better solution of the Boltzmann equation were implemented.

LSMS method described above, it is possible to calculate the electronic structure for supercells containing thousands of atoms with no approximation other than the DFT. In figure 5.13 we show the charges on 512 copper atoms and 512 zinc atoms distributed randomly on the sites of a supercell with the structure of a body-centered cubic lattice. The distribution is broad, and upon reflection is what would be expected on the basis of physical intuition. It should be compared with the prediction of the isomorphous model that all the Cu atoms would take on exactly 0.09978 electron charges, and Zn atoms would lose the same amount.

A version of the CPA can be constructed that leads to a polymorphous model of the alloy in which each atom has a different charge, [17] but it requires a different kind of averaging [18]. Recall that we consider periodically reproduced supercells. In the binary alloy model, N_A A atoms and N_B B atoms are distributed over the $N = N_A + N_B$ sites. The Green's function defined in equation (4.64) for the alloy is different for each of the $M_c = N!/N_A!N_B!$ possible atomic arrangements, and can be labeled $G_n^i(E, \mathbf{r}_n, \mathbf{r}'_n)$ to emphasize that. Then, the ensemble average in equation (5.3) can more properly be written

$$\left\langle G_n(E, \mathbf{r}_n, \mathbf{r}'_n)\right\rangle = \frac{1}{M_c}\sum_{i=1}^{M_c} G_n^i(E, \mathbf{r}_n, \mathbf{r}'_n). \tag{5.32}$$

It is also possible to consider one supercell with the atomic arrangement i, and average over the sites

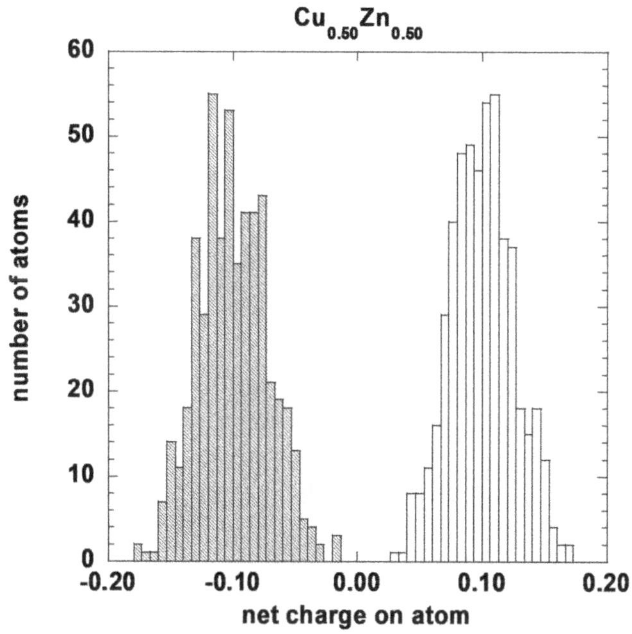

Figure 5.13. The distribution of charges on the sites of a 50% copper–zinc alloy modeled with a supercell containing 1024 atoms. The charges on the copper sites are shown with the white histograms, and the charges on the zinc sites are shown in black.

$$\left\{G^i(E, \mathbf{r}, \mathbf{r}')\right\} = \frac{1}{N}\sum_{n=1}^{N} G_n^i(E, \mathbf{r}, \mathbf{r}'). \tag{5.33}$$

We have seen that, as $N \to \infty$, $\langle G_n(E, \mathbf{r}_n, \mathbf{r}'_n)\rangle$ is independent of the site index n, it is equally clear that in the same limit, $\{G^i(E, \mathbf{r}, \mathbf{r}')\}$ is independent of the arrangement index i. The equivalence of the ensemble average in equation (5.32) with the site average in equation (5.33) is similar to the Ergodic theorem first proposed by Boltzmann, and can be proved by similar arguments. If the condition is imposed that the alloy must be isomorphous, the site average will lead to the same averaged potential as the ensemble average result shown in equation (5.3). However, the site average can lead to a polymorphous model.

Applying the CPA philosophy to the site averaged Green's function leads to

$$\{G(E, \mathbf{r}, \mathbf{r}')\} = \frac{1}{N}\sum_{n=1}^{N}\{G_n(E, \mathbf{r}, \mathbf{r}')\}, \tag{5.34}$$

where

$$\{G_n(E, \mathbf{r}, \mathbf{r}')\} = \sum_{LL'} Z_L^n(E, \mathbf{r})\left\{\tau_{LL'}^{nn}(E)\right\}Z_{L'}^n(E, \mathbf{r}') - \sum_L Z_L^n(E, \mathbf{r})J_L^n(E, \mathbf{r}'), \tag{5.35}$$

and

$$\{\boldsymbol{\tau}^{nn}\} = \left[\mathbf{I} - \boldsymbol{\tau}_c^{nn}(\mathbf{m}_n - \mathbf{m}_c)\right]^{-1}\boldsymbol{\tau}_c^{nn}. \tag{5.36}$$

Inserting the Green's function for an index n into this equation produces a charge density, $\rho_n(\mathbf{r})$, and hence a one-electron potential, $v_n(\mathbf{r})$, for every site. From the potential, we calculate the t-matrix $\mathbf{t}_n$ and its inverse $\mathbf{m}_n$. As in the isomorphous CPA, the scattering path for the coherent potential host lattice $\boldsymbol{\tau}_c^{nn}$ is calculated from equation (5.8). The condition that defines the coherent potential is

$$\frac{1}{N}\sum_{n=1}^{N}\{\tau^{nn}\} = \tau_c^{nn}, \tag{5.37}$$

instead of equation (5.11).

Since we are focusing on a specific supercell, the position $\mathbf{R}_n$ of each atom is known and the Madelung potential for the nth site is

$$v_n^M = \sum_{m\neq n=1}^{N} M(|\mathbf{R}_n - \mathbf{R}_m|)q_n, \tag{5.38}$$

where

$$q_n = \int_{\Omega_n}\rho_n(\mathbf{r})d\mathbf{r} - Z_n, \tag{5.39}$$

and $M(|\mathbf{R}_n - \mathbf{R}_m|)$ is the Madelung matrix. The Madelung matrix is defined so that the sum gives the contribution to the Coulomb potential from all of the periodically reproduced supercells. A sophisticated method for calculating it was developed by P P Ewald, and it reduces to $1/|\mathbf{R}_n - \mathbf{R}_m|$ when $N \to \infty$. The potentials $v_n(\mathbf{r})$ contain the Madelung shift v_n^M.

The polymorphous version of the CPA (PCPA) has been applied to several alloy systems. For alloys that have a large amount of charge transfer, such as copper–gold or copper–palladium, better results are given than the ones obtained with the isomorphous CPA, but the improvements are quite small for alloys like silver–palladium that have a small charge transfer [19]. The polymorphous CPA has the disadvantage that it requires the use of a specific supercell. This is a small price to pay for the improvement in the treatment of the Coulomb energy. It is quite easy to apply because the charge densities $\rho_A(\mathbf{r})$ and $\rho_B(\mathbf{r})$ and the effective t-matrix $\mathbf{t}_c$ from the isomorphous CPA are used as the starting values in the self-consistent iterations.

Although the coherent potential $v_c(\mathbf{r})$ is introduced into the discussion, it is never calculated or used. The only potentials used to calculate wave functions are real potentials for real atoms, $v_n(\mathbf{r})$. From the analytic properties of $\mathbf{t}_c$ it follows that $v_c(\mathbf{r})$, if it existed, would be complex and energy-dependent.

CPAs, either the KKR-CPA or the PCPA, have some advantages. It is useful on occasion to know if a given effect can be explained at the level of a mean-field theory. The approximation that there is an effective scatterer $\mathbf{t}_c$ can be used in the development of other theories. For that reason, we did a study comparing calculated heats of mixing in Cu–Zn alloys with four different concentrations. The results are shown in figure 5.14. As might be expected, the best agreement with experiment is

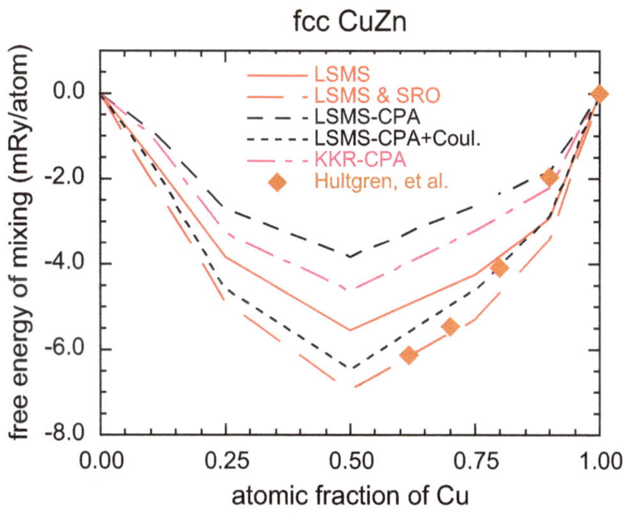

Figure 5.14. The energy of mixing calculated with the order-N LSMS method for no short-range order (solid line) and an estimate of the experimental short-range order (the line with long dashes). The calculations were carried out for fcc copper and zinc and for models of 10%, 25%, 50%, 75%, and 90% fcc copper–zinc alloys with supercells that contain 256 atoms. The lattice constants were obtained from LSMS calculations. The LSMS-CPA (PCPA) and KKR-CPA energies of mixing are shown with the line with long dashes and the line with long and short dashes. The LSMS-CPA plus the interatomic Coulomb energy is shown with the dashed line. The experimental energies of mixing are shown with the diamond-shaped marks.

obtained with the LSMS, particularly if the samples are modified to show the experimental short-range order. The KKR-CPA is slightly closer to experiment than the PCPA, but neither are very close. Simply adding the Coulomb energy to the LSMS-CPA (PCPA) gets remarkably good agreement.

5.8 Historical studies of alloys

The first systematic studies of alloys were carried out by the metallurgist William Hume-Rothery. He developed a famous set of rules that describe the conditions under which an element could dissolve in a metal, forming a solid solution. One set of rules refers to substitutional solid solutions, and the other refers to interstitial solid solutions. He also collaborated with the theoretical physicists, H Jones and N F Mott, on predictions of the boundaries of the various structural phases in the equilibrium phase diagram of alloys, which are described in a famous book by these two physicists [20].

Perhaps the best example of an alloy system described by Mott and Jones as a Hume-Rothery system is copper–zinc. Pure copper has a face-centered cubic (fcc) structure. When zinc atoms are substituted for copper atoms to form a substitutional solid solution, the structure of the resulting alloys remains fcc until the atomic percentage of copper atoms reaches 38.4%. Alloys with a slightly higher concentration of zinc are a mixture of two structural phases. There are regions of the material that have a fcc phase, and other regions that have a body-centered cubic (bcc) structure. Alloys with higher zinc concentrations are in the pure bcc phase, but when the concentration becomes higher than 48% a complicated structure called the γ phase begins to appear.

In the 1930s, when Jones began to address the question of why these alloys changed their structure at the specific concentrations described above, it was understood that many properties of metals could be explained by applying the newly developed quantum mechanics to a gas of almost free electrons. In particular, Sommerfeld and his collaborators realized that the statistical behavior of the electron gas is described by the Fermi–Dirac distribution. According to the exclusion principle, only two electrons could be put into a state with pseudomenum $\hbar\mathbf{k}$, one with spin up and the other with spin down. When all the lowest energy states are filled, the k-vectors for the states fill a sphere, called the Fermi sphere (FS), with radius $k_F = (3\pi^2 e/a)^{1/3}$, where e/a is the number of electrons per atom. In the free electron theory of a solid, there is an additional condition on the k-vectors. They must also fit within a unit cell in k space called the Brillouin zone (BZ).

The number of electrons per atom were taken from the known chemical data, so copper was assumed to have one electron per atom and zinc to have two. For pure copper, e/a is one and the FS fills one half the volume of the BZ. As zinc is added to the system, e/a increases, the FS becomes larger, and the electronic density of states of the alloys increases like the square root of the Fermi energy, $E_F = \hbar^2/2mk_F^2$. For an alloy with 36.2% zinc, the FS touches the octagonal face of the fcc BZ at the point called L in figure 5.15. In alloys with a higher concentration of zinc, k_F is greater than the distance from zero to L, k_{crit}^{fcc}. Portions of the FS will then lie outside the BZ, and the electronic density of states begins to decrease. According to this analysis, the electronic density of states of pure fcc zinc should be zero because the surface of the FS is completely outside the BZ. Of course, pure zinc does not have a fcc structure.

A simple thermodynamic argument shows that the total energy of the alloy can be reduced by transforming from the fcc to the bcc crystal structure when $e/a \geqslant 1.362$ and $k_F \geqslant k_{crit}^{fcc}$. The bcc alloy will become unstable when $e/a \geqslant 1.480$ because k_F is larger than k_{crit}^{bcc}, the distance from 0 to the center of the rhomboidal face marked N in figure 5.15. It is energetically favorable for alloys with higher zinc content to transform the γ phase mentioned above.

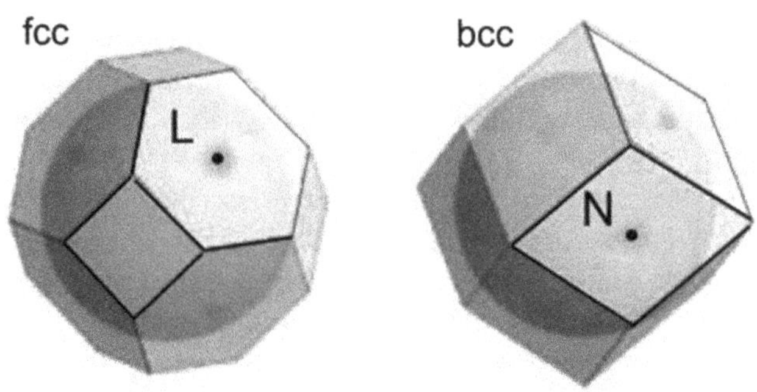

Figure 5.15. Brillouin zones for face-centered and body-centered cubic lattices and inscribed spheres.

According to table VI of chapter 5 of reference [20], the *e*/*a* values at which the transformations take place are 1.384 and 1.480. Other alloy systems are shown in the table. Taking the *e*/*a* of pure aluminum to be 3, the transformations for copper aluminum alloys occur at 1.408 and 1.480. Similar results are shown for alloys based on the other noble metals, silver and gold.

Today, hundreds of alloy systems that fall into the category of noble metals alloyed with higher valence transition metals, post transition metals, and metalloids have been studied. The rules outlined by Mott and Jones predict the boundaries in the phase diagrams with reasonable accuracy. With all of this success, it would seem that the underlying theory must be correct. It thus came as a surprise when later calculations showed that the underlying assumptions are not borne out.

As was pointed out above, one of the unique features of the KKR band theory method is that it is as easy or even easier to search for zeros of the KKR matrix, equation (3.58), by fixing the energy and varying the k-vector. It is known that the symmetry of fcc crystals like copper is such that the complete constant energy surface can be found by applying rotation operators to the surface in one forty-eighth of the Brillouin zone. In one of the first KKR calculations [21], constant energy surfaces were calculated for twenty-five energies in the neighborhood of the Fermi energy. For each energy, the surface is defined by 561 radii in k space. That means that there are 26 066 k points on the entire constant energy surface. The volume of k space bounded by the constant energy surface gives the number of eigenstates with energy less than E, which is called the integrated density of states $M(E)$. The most accurate calculation of the density of states $\rho(E)$ in copper that had been done at that time was done by numerically differentiating $M(E)$. This is shown in figure 5.16.

The theoretical model used by Mott and Jones predicts that the density of states should increase smoothly like the square root of E throughout the energy region shown in this figure. The constant energy surfaces do not touch the surface of the Brillouin zone until much higher energies. The Fermi surface predicted by the calculations of reference [14] (chapter 1) agrees with the de Haas van Alphen measurements, [22] which show a neck around the points L.

The data in figure 5.16 are in radical disagreement with the assumptions of Mott and Jones, but the story doesn't end there. It turns out that the measured density of states increases when zinc is alloyed into copper [23]. The density of states in metals is obtained by fitting the low temperature specific heat to the formula

$$C_p(T) = \gamma T + bT^3, \tag{5.40}$$

where it was originally thought that γ is proportional to the calculated density of states

$$\hat{\gamma} = 1/3\pi^2 k_B^2 \rho(E_F). \tag{5.41}$$

If this were the case and the rigid band model were applied to the data in figure 5.16, $\hat{\gamma}$ would decrease with increasing zinc concentration with the slope shown in that

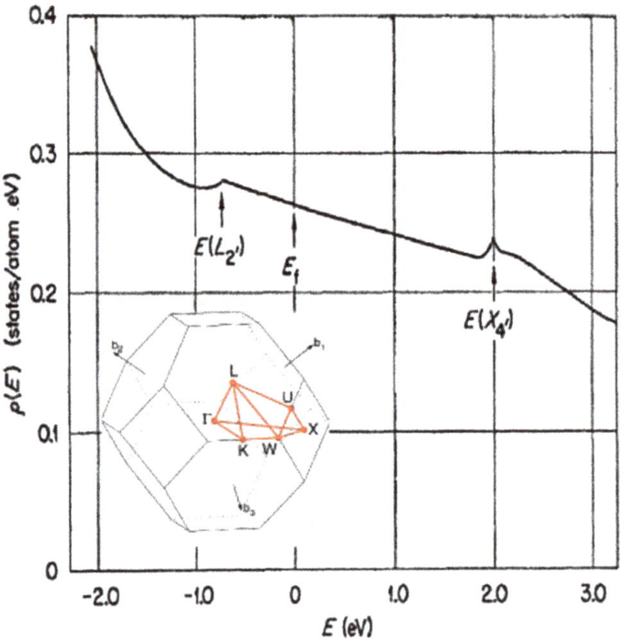

Figure 5.16. The density of states of copper from 2.0 eV below the Fermi energy to 3.0 eV above. More energies were used in the neighborhood of the peaks than in the smooth regions. The Fermi energy is indicated by the E_f. The fcc Brillouin zone is inserted, and the symmetry points of the fundamental 1/48th of the zone are shown. The peak marked $E(L_{2'})$ arises when the constant energy surface touches the octagonal face at the symmetry point L. The one at $E(X_{4'})$ is at the energy for which it touches at the point X. Reprinted with permission from [21]. ©1967 American Physical Society.

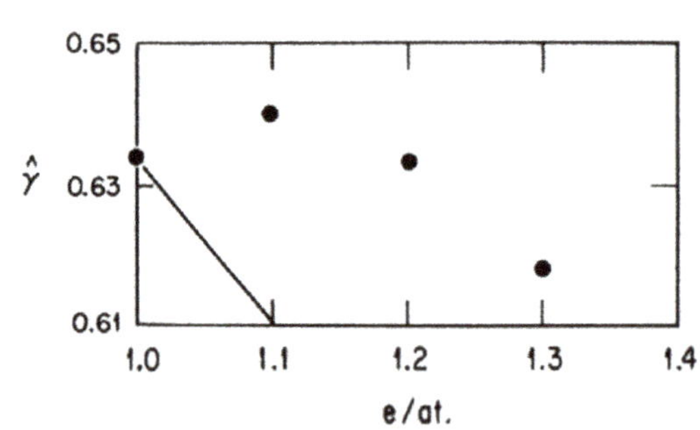

Figure 5.17. The constants $\hat{\gamma}$ calculated with the density of states calculated for pure copper and for alloys with 10%, 20%, and 30% zinc are shown by dots. The straight line shows $\hat{\gamma}$ with the densities of states obtained by applying the rigid band model to the data from the copper calculation. Reprinted with permission from [24]. ©1981 American Physical Society.

figure. Since that does not agree with the experiment, $\rho(E_F)$ is obtained from CPA calculations on this alloy system [24] (figure 5.17 shows the results). It is clear that the rigid band model cannot be used, but the values of γ still do not agree with the experiment.

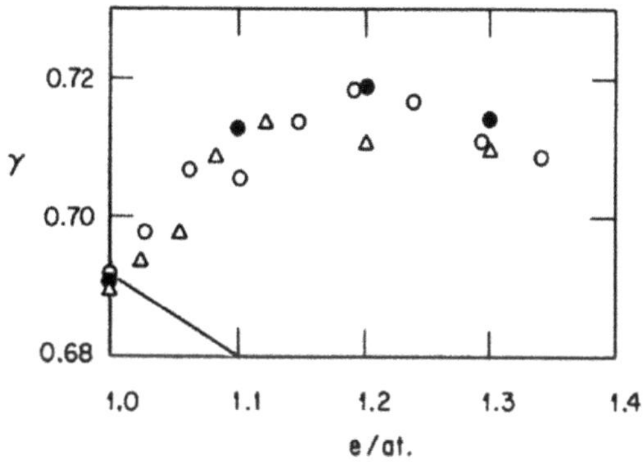

Figure 5.18. The black dots show the enhanced coefficient γ calculated from the theory. The open circles and triangles are the experimental data from reference [23]. Reprinted with permission from [23]. © 1981 American Physical Society.

Many body theorists demonstrated that the coefficient is enhanced by electron–phonon interactions [25]. The observed value is given by

$$\gamma = \hat{\gamma}(1 + \lambda), \tag{5.42}$$

where the dimensionless enhancement factor is roughly 0.1. Using this value, theoretical predictions are compared with the experimental results in figure 5.18.

The result of this analysis is the proposition that the properties of metals and alloys can be interpreted by applying quantum mechanics to the electrons. However, the true explanation may well require extensive calculations requiring a great deal of computer power.

References

[1] Wang Y, Stocks G M, Shelton W A, Nicholson D M, Szotek Z and Temmerman W M 1995 *Phys. Rev. Lett.* **75** 2867
[2] Faulkner J S, Wang Y and Stocks G M 1998 *Phys. Rev. Lett.* **81** 1905
[3] Faulkner J S, Wang Y and Stocks G M 1995 *Phys. Rev.* B **52** 17106
[4] Gyorffy B L *et al* 1985 *J. Phys.* F **15** 1337
[5] Staunton J B and Gyorffy B L 1992 *Phys. Rev. Lett.* **69** 371
[6] Wang Y, Stocks G M and Nicholson D M C *et al* 1997 *J. Appl. Phys.* **81** 3873
[7] Stocks G M *et al* 1998 *Phil. Mag.* B **78** 665
[8] Antropov V P *et al* 1996 *Phys. Rev.* B **54** 1019
[9] Faulkner J S and Stocks G M 1980 *Phys. Rev.* B **8** 3222
[10] Lloyd P 1967 *Proc. Phys. Soc.* **90** 207
[11] Stocks G M, Williams R W and Faulkner J S 1971 *Phys. Rev.* B **4** 4390
[12] Gyorffy B L, Stocks G M and de Physique J 1974 Colloque C4 *Tome* **35** C4
Stocks G M, Temmerman W M and Gyroffy B L 1978 *Phys. Rev. Lett.* **41** 339
[13] Pinski F J, Allen P B and Butler W H 1978 *Phys. Rev. Lett.* **41** 431
[14] Gyorffy B L and Stocks G M 1983 *Phys. Rev. Lett.* **50** 374
[15] Staunton J, Gyorffy B L, Stocks G M and Wadsworth J 1986 *J. Phys.* F **16** 761

[16] Butler W H and Stocks G M 1984 *Phys. Rev.* B **29** 4217
[17] Ujfalussy B, Faulkner J S, Moghadam N Y, Stocks G M and Wang Y 2000 *Phys. Rev.* B **61** 12005
[18] Faulkner J S 2001 *Phys. Rev.* B **64** 233113
[19] Pella S, Faulkner J S and Malcolm G 2004 Stocks, and Balazs Ujfalussy *Phys. Rev.* B **70** 064203
[20] Mott N F and Jones H 1936 *The Theory of the Properties of Metals and Alloys* (Oxford: Clarendon Press)
[21] Faulkner J S, Davis H L and Joy H W 1967 *Phys. Rev.* **161** 656
[22] Shoenberg D and Trans P 1962 *R. Soc. London* **255** 85
[23] Veal B W and Rayne J A 1963 *Phys. Rev.* **130** 2156
Mizutani U, Noguchi S and Massalski T B 1972 *Phys. Rev.* B **5** 2057
[24] Faulkner J S and Stocks G M 1981 *Phys. Rev.* B **23** 5628
[25] McMillan W L 1968 *Phys. Rev.* **167** 331

IOP Publishing

Multiple Scattering Theory
Electronic structure of solids
J S Faulkner, G Malcolm Stocks and Yang Wang

Chapter 6

Spectral theory in multiple scattering theory

To this point we have been outlining a formalism that makes it possible to solve the Schrödinger equation for condensed matter in its entirety. That is, all of the wave functions, Green's functions, and energy eigenvalues are obtained. Many years ago, mathematicians found that a great deal of information about the spectrum of an operator can be obtained without solving for all aspects of the problem. In the present context, the term spectrum refers to the distribution of the energy eigenvalues of the system under investigation.

Physicists have independently come upon relationships that fall in the category of spectral theory. One of the oldest of these is the Friedel sum

$$N(E) = \frac{2}{\pi}\sum_{l=0}^{l_{\max}}(2l + 1)\delta_l(E). \tag{6.1}$$

In this expression, $N(E)$ is the integrated density of states (IDOS) associated with a single atom embedded in a vacuum (or an impurity atom in an otherwise ordered solid.) The $\delta_l(E)$ are the phase shifts that describe the scattering by the atom, and are defined in equation (2.64). The number of core states of this atom is given by Levinson's theorem

$$n_{\text{core}}(E) = \frac{2}{\pi}\sum_{l=0}^{l_{\max}}(2l + 1)\delta_l(0), \tag{6.2}$$

where

$$\delta_l(0) = \operatorname*{Lim}_{E\to 0} \delta_l(E). \tag{6.3}$$

Another formula in this category is Lloyd's formula [1] that relates the (IDOS) of the collection of atoms in a solid to the scattering matrix $\mathbf{M}(E)$ defined in equation (3.35), or the integral over $\mathbf{k}$ of the KKR matrix from equation (3.56)

doi:10.1088/2053-2563/aae7d8ch6

$$N(E) - N_0(E) = -\frac{1}{\pi}\text{Im}\ln\det\mathbf{M}(E). \tag{6.4}$$

The function $N_0(E)$ is the IDOS for a free-electron gas.

We have already discussed that the density of states for one or more atoms can be obtained from the Green's function for the system

$$n(E) = \frac{dN(E)}{dE} = -\frac{2}{\pi}\text{Im}\int_{\substack{\text{all}\\ \text{space}}} G(E, \mathbf{r}, \mathbf{r})d\mathbf{r}, \tag{6.5}$$

as shown in equation (4.5). This can also be considered a spectral equation.

Although these formulae were arrived at in very different ways, a great deal of insight can be gained about them by relating them to a theorem proved by the Ukranian mathematician Mark Grigorievich Krein [2]. In the 1930s, Krein created one of the strongest centers of functional analysis in the world at Odessa University. Many of the results of this period by him, as well as in collaborations with his friends, colleagues, and students are now characterized as classical and appear in textbooks on functional analysis.

6.1 Krein's theorem

Krein proved that, for any two self-adjoint operators whose difference is trace class, there exists a real function of real variables $\xi(E)$ such that

$$Tr[\phi(H) - \phi(H_0)] = \int_{-\infty}^{\infty}\frac{d\phi(x)}{dx}\xi(x)dx, \tag{6.6}$$

where $\phi(X)$ is any sufficiently smooth function. The notation $Tr[X]$ designates a process that is the abstract version of the trace operation. For applications to scattering theory, the self-adjoint operators are chosen to be the Hamiltonian operator for the system H and for a free particle H_0. The potential, $V = H - H_0$ is in the trace class for most cases of physical interest. Also, the arbitrary smooth function is chosen to be $\phi(H) = (z - H)^{-1}$, where z is a complex number. We choose to integrate over the energy E, which means that $\phi(E) = (z - E)^{-1}$. It follows that

$$Tr[(z - H)^{-1} - (z - H_0)^{-1}] = \int_{-\infty}^{\infty}\frac{\xi(E')}{(z - E')^2}dE'. \tag{6.7}$$

Birman and Krein [3] proved that Krein's spectral shift function $\xi(E)$ is given by

$$e^{-i2\pi\xi(E)} = \det\mathbf{S}(E), \tag{6.8}$$

where $\mathbf{S}(E)$ is the standard unitary S-matrix. The uniqueness of $\xi(E)$ is demonstrated in the papers quoted, and in the massive literature on the subject that followed. In simple terms, the definition includes the conditions that $\xi(E)$ is a continuous function of E and approaches zero as E approaches plus or minus infinity. The spectral shift function defined in this way is consistent with the known properties of the S-matrix. From this definition and Cauchy's integral theorem, it follows that

$$\operatorname*{Lim}_{z \downarrow E} \int_{-\infty}^{\infty} \frac{\xi(E')}{(z - E')^2} dE' = i\pi \frac{d\xi(E)}{dE}. \tag{6.9}$$

We want to apply Krein's theorem to the case of a single potential embedded in free space. Recall that the potentials of interest in MST are such that $v(\mathbf{r})$ is defined to be zero for $\mathbf{r}$ outside the finite domain Ω, as discussed in chapter 2. It follows that the energy spectrum for a system with one such potential embedded in a vacuum is discrete for $E < 0$, being the set of bound state energies E_v. The system has a continuous spectrum for $E \geqslant 0$.

The trace operation in equation (6.7) is carried out by taking the integral over all space, which leads to

$$Tr[(z - H)^{-1} - (z - H_0)^{-1}] = \int_{\infty} [G(E, \mathbf{r}, \mathbf{r}) - G_0(E, \mathbf{r}, \mathbf{r})] d\mathbf{r}, \tag{6.10}$$

where $G(E, \mathbf{r}, \mathbf{r})$ is the diagonal element of the Green's function defined in equation (4.3). We know from the discussion in chapter 4 that the integral will give the sum of all the eigenvalues of the Hamiltonian H minus the eigenvalues of H_0

$$\left[2\sum_v d_v \delta(E - E_v) + [n(E) - n_0(E)] \right] = -2\frac{d\xi(E)}{dE}, \tag{6.11}$$

where d_v is the degeneracy of the bound state with energy E_v, $n(E)$ is the density of states, and $n_0(E)$ is the free-electron density of states. From the properties of the potential, $v(\mathbf{r})$, we know that $n(E)$ and $n_0(E)$ are zero for $E < 0$ and that all the bound state energies are negative. The '2' is included to account for the spin of the electrons. Integrating equation (6.11) from minus infinity to E, with $E \geqslant 0$, leads to

$$-2\xi(E) = N(E) - N_0(E) + n_c = N_K(E), \tag{6.12}$$

where n_c is the number of core electrons. The function $N_K(E)$ is called the Krein integrated density of states (IDOS).

Integrating equation (6.11) from minus infinity to zero leads to a generalization of Levinson's theorem

$$2\sum_v d_v = n_c = -2\xi(0) = N_K(0). \tag{6.13}$$

This equation, which holds for potentials that are not spherically symmetric, can be checked for consistency with the ordinary Levinson's theorem by applying it to potentials that are.

It is obvious that the advantage of Krein's formulation is its generality. The trace operation in equations (3.1) and (3.2) is independent of the orthonormal basis used, as long as it is complete. The versions familiar in quantum theory are spatial integrations, as in equation (2.1), or integrations over momenta. The requirement that the potential is trace class is essentially equivalent to the condition that the trace on the left side of equation (6.7) exists, with no shape approximation. As with the

Green's function formulae in the preceding section, the ones based on Krein's theorem are made tractable by ignoring contributions from terms with $l > l_{max}$.

Krein's theorem, as stated in equation (6.7), holds for any energy. Calculations in the range $-\infty < E < 0$ are very difficult for the specific kind of potentials being treated in this work. For this reason, a numerical test of the generalized Levinson theorem, equation (6.13), for these potentials is carried out by calculating $\xi(E)$ for $0 \leqslant E < E_{max}$, where E_{max} is sufficiently large that $\xi(E)$ has become a constant. It can be shown that

$$\xi(E) \to 0 \text{ as } E \to \infty, \tag{6.14}$$

so $\xi(E_{max})$ is set equal to zero, and integers are added to $\xi(E)$ over various energy intervals in such a way as to make it a continuous function. Following the function in this way from E_{max} to zero is a way of counting the bound states of the potential using equation (6.16). The result can be checked by finding the bound states with the numerical techniques that have historically been used in atomic physics.

The S-matrix for the special case of a spherically symmetric potential is diagonal with elements

$$s_l(E) = e^{2i\delta_l(E)}, \tag{6.15}$$

repeated $2l + 1$ times. Considering only the states that correspond to a given l, equation (6.13) is equivalent to

$$n_c^{(l)} = -2\xi_l(0) = \frac{2}{\pi}\delta_l(0), \tag{6.16}$$

the standard version of Levinson's theorem.

It is also clear from equation (6.15) that Krein's theorem for a spherical potential provides a mathematical justification for the Friedel sum, equation (6.1). The Friedel sum has been used for many years as a more or less ad hoc method for estimating the density of states (DOS) or the IDOS for an impurity embedded in a free-electron solid. This connection with Krein's theorem, and the detailed study in the next section of the connection between the DOS from Krein's theorem and the one that is obtained by taking the trace of the Green's function, clarifies the meaning of the Friedel sum far beyond most of the derivations in the literature. To be specific, the impression is usually given that the change in the DOS caused by the potential $v(\mathbf{r})$ is restricted to the region of space within which $v(\mathbf{r}) \neq 0$. In the next section we find that this is far from the case. The Green's function $G(E, \mathbf{r}, \mathbf{r})$ must be integrated over values of $\mathbf{r}$ that extend all the way to infinity in order to produce a DOS that is precisely equal to the one given by Krein's theorem. The interpretation of this, which becomes less surprising after reflection, is that a potential with a finite domain can influence states in all of space.

6.2 Calculations with real potentials using Krein's theorem

At this point, the way that Krein's spectral displacement function compares with densities of states and integrated densities of states obtained from Green's functions is understood. The crucial question is if the theory is applicable to full potential multiple scattering calculations, and if anything of value for such applications has been learned. This question was addressed by the authors of the present book, and published in a special issue of the *Journal of Physics: Condensed Matter*, focusing on Correlation and Disorder [4]. The collection of papers was published to commemorate the life of Professor Balazs Gyorffy.

The single site potentials used in this section were extracted from self-consistent full potential LSMS calculations on periodic crystals of aluminum, copper, and molybdenum. The lattice constants for the face-centered cubic crystals of Al and Cu were 7.65 and 6.76 Bohr, while for body-centered cubic Mo it was 5.913 Bohr. To obtain potentials suitable for the present purposes we used minimal settings of the usual l-truncation parameters, namely $l_{\max\,\mathrm{kkr}} = 2$, $l_{\max\,\mathrm{phi}} = 2$, $l_{\max\,\mathrm{rho}} = 4$, $l_{\max\,\mathrm{pot}} = 4$, and $l_{\max\,\mathrm{truc}} = 12$, where $l_{\max\,\mathrm{kkr}}$, $l_{\max\,\mathrm{phi}}$, $l_{\max\,\mathrm{rho}}$, $l_{\max\,\mathrm{pot}}$, and $l_{\max\,\mathrm{truc}}$ are the angular momentum expansion cut-offs for the KKR matrix, wave-function, charge density, single site potential and $\sigma(\mathbf{r})$-function, respectively (the latter is used when performing spatial integrations over a Voronoi polyhedron). The von Barth-Hedin exchange-correlation functional was used in all of the calculations. Because Al, Cu, and Mo are all cubic crystals the output potentials contain both a spherical, $l = 0$, and non-spherical, $l = 4$, components. Of course, a fully converged full potential calculation requires much larger values for these parameters. For studying the single site Green's function and Krein's theorem we use $l_{\max\,\mathrm{kkr}} = 8$ and $l_{\max\,\mathrm{truc}} = 12$ together with the above $l_{\max\,\mathrm{pot}} = 4$ single site potential. When solving the single site equations we further apply the $\sigma(\mathbf{r})$-function to the untruncated potential, terminating this expansion at $l = 8$. As a consequence, the actual potential used in solving the single site equation contains $l = 0$, 4, 6 and 8 components. For performing the energy integral we tested a variety of linear and logarithmic grids with different lower limits for the integration within the range 10^{-8} to 10^{-3} Ry (the upper limit being set at 1.5 Ry) to test overall convergence of the various integrals. To test the small-energy asymptotic behavior of various partial quantities and the generalized phase we also used dense (~1000-point) linear and logarithmic meshes with lower/upper limits of $10^{-8}/10^{-3}$ Ry; the results of these exercises were consistent with what was already expected from previous studies of the square well (above) and muffin-tin potentials. The specific full potential results shown below are for a logarithmic mesh with 1000-steps and a lower limit of 10^{-3} Ry, which are sufficient for the present purposes.

In FP calculations it is obviously necessary to use the general formula for the S-matrix

$$\mathbf{S}(E) = \mathbf{I} - i2\alpha\mathbf{t} = (\mathbf{c} + i\mathbf{s})(\mathbf{c} - i\mathbf{s})^{-1}, \tag{6.17}$$

in the Birman–Krein formula equation (6.8) for the spectral displacement function. This matrix can easily be seen to be normal and unitary. Next, it has been found that it is better to replace equation (4.51) with

$$n_{\mathrm{in}}(E) = \frac{2\alpha}{\pi}\mathrm{Im}\sum_{L}\sum_{L'}\int_{\Omega}\phi_L(E, \mathbf{r})\left[\tilde{\mathbf{s}}^{\bullet}(i\mathbf{s} - \mathbf{c})\right]^{-1}_{LL'}\phi^{\bullet}_{L'}(E, \mathbf{r})d\mathbf{r}, \tag{6.18}$$

when calculating the contribution to $n(E)$ from inside the Voronoi polyhedron Ω. The two formulae are mathematically equivalent, but equation (6.18) gets around the problem that the $\tilde{\mathbf{c}}\mathbf{c}$ term in Ξ is badly behaved for large $l_{\max}$.

The portion of $n(E)$obtained by integrating the Green's function over the space outside the Voronoi polyhedron, $\Omega_\infty - \Omega$, is called $n_{\text{out}}(E)$. To calculate this quantity, we consider two concentric spheres. The largest sphere that can be inscribed within Ω has a radius R_{mt}, while the smallest sphere that circumscribes Ω has a radius R_c. We introduce a step function $\sigma(\mathbf{r})$ that is one inside the Voronoi polyhedron and zero outside. The technique for expanding this function

$$\sigma_{LL'}(r) = \int_{4\pi} \sigma(\mathbf{r}) Y_L(\mathbf{r}) Y^*_{L'}(\mathbf{r}) d\mathbf{r}, \tag{6.19}$$

has been well established [5]. Using these quantities we obtain

$$\begin{aligned} n_{\text{out}}(E) = &\frac{2\alpha^2}{\pi} \text{Im} \sum_L t_{LL}(E) \int_{R_{mt}}^{\infty} h_l(\alpha r)^2 r^2 dr \\ &- \frac{2\alpha^2}{\pi} \text{Im} \sum_{L,L'} t_{LL'}(E) \int_{R_{mt}}^{R_c} [h_l(\alpha r) \sigma_{LL'}(r) h_{l'}(\alpha r)] r^2 dr. \end{aligned} \tag{6.20}$$

This advantage in writing $n_{\text{out}}(E)$ in this form is that the definite integral in this equation has been worked out and published [6]

$$\int_R^{\infty} h_l(\alpha r)^2 r^2 dr = \frac{R^3}{2} \left[h_{l-1}(\alpha R) h_{l+1}(\alpha R) - h_l(\alpha R)^2 \right]. \tag{6.21}$$

Using these formulae, the partial densities of states for the Al, Cu, and Mo potentials were calculated at 1000 energy points on a logarithmic scale between $E_{\min} = 10^{-3}$Ry and $E_{\max} = 1.5$Ry. The partial IDOS's, $N_{\text{in}}(E)$ and $N_{\text{out}}(E)$, were obtained by numerical integration. The equality

$$N_K(E) = N_{\text{in}}(E) + N_{\text{out}}(E), \tag{6.22}$$

appears to be the case from the curves shown in figures 6.1–6.3. The actual numbers demonstrate this even more convincingly because the equality holds to the accuracy of the computer.

The densities of states for the three elements are shown in figure 6.1 The solid line shows the Krein DOS $n_K(E)$ for aluminum. The dashes and the dash-dot lines show the partial DOS's $n_{\text{in}}(E)$ and $n_{\text{out}}(E)$ calculated using the Green's function for the same element.. The Krein DOS $n_K(E)$ was obtained by numerically differentiating the IDOS $N_K(E)$.

The Krein DOS is the sum of those calculated from the Green's function, as expected. In addition, $n_K(E)$ and $n_{\text{out}}(E)$ approach infinity as E approaches zero. The simple proof of this for spherical potentials is that the $l = 0$ component of the IDOS can be shown to behave like $N_0(E) \propto \sqrt{E}$ for small energy. It follows that $n_0(E) \propto E^{-1/2}$ for $E \approx 0$. It is hard to show this for the Krein IDOS, but an argument

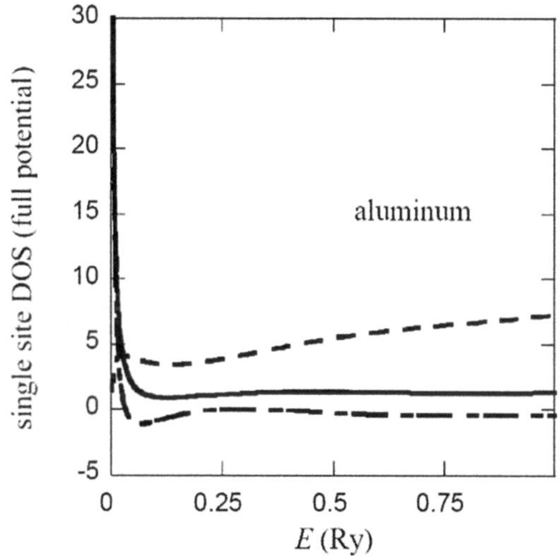

Figure 6.1. The solid line shows the Krein DOS $n_K(E)$ for aluminum. The dashes and the dash-dot lines show the partial DOS's $n_{\text{in}}(E)$ and $n_{\text{out}}(E)$ calculated using the Green's function for the same element. Reproduced from [4].

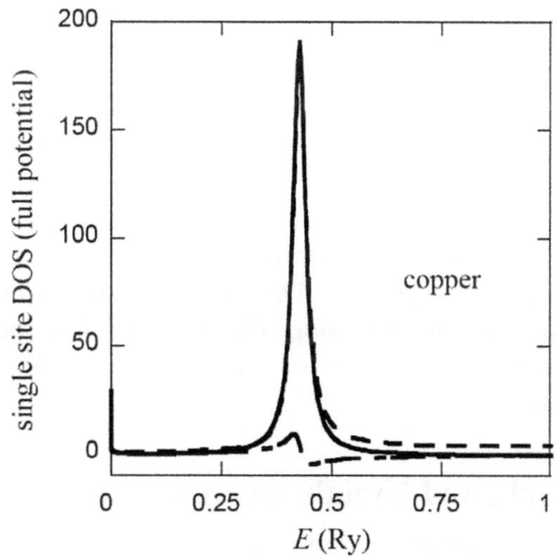

Figure 6.2. The solid line shows the Krein DOS $n_K(E)$ for copper. The dashes and the dash-dot lines show the partial DOS's $n_{\text{in}}(E)$ and $n_{\text{out}}(E)$ calculated using the Green's function for the same element. Reproduced from [4].

first introduced by Gyorffy [7] is helpful in this connection. Because the S-matrix is normal and unitary, it has eigenvalues $\lambda_n(E)$ and they fall on a unit circle, i.e.,

$$\lambda_n(E) = e^{i2\vartheta_n(E)} \tag{6.23}$$

with $\vartheta_n(E)$ real numbers. Gyorffy coined the term generalized phase shifts for these functions. The generalized phase shifts have been calculated for full potential

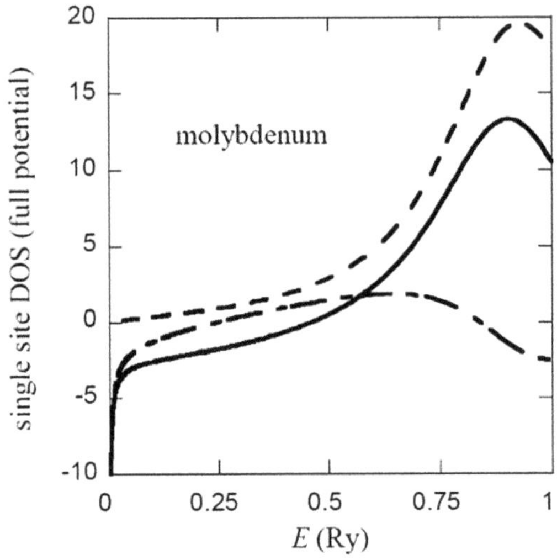

Figure 6.3. The solid line shows the Krein DOS $n_K(E)$ for molybdenum. The dashes and the dash-dot lines show the partial DOS's $n_{\text{in}}(E)$ and $n_{\text{out}}(E)$ calculated using the Green's function for the same element. Reproduced from [4].

aluminum, copper, and molybdenum. At small energies they are dominated by an *s*-like phase shift that is proportional to the square root of E. The other phase shifts depend on energy but there are more than one for each l. This is interesting in itself, and it provides the explanation for the form of the Krein spectral displacement function of full potentials.

Another observation is that the IDOS and DOS of copper are dominated by the portion that comes from inside the Voronoi polyhedron, which is not surprising for an element whose behavior is dominated by localized d-functions. For aluminum, an almost free-electron material, the opposite is true. The early transition-metal molybdenum falls in between. These assumptions could only be substantiated by doing the calculations.

6.3 Lloyd's formula and Krein's theorem

The Lloyd's formula for a cluster of atoms is

$$N(E) - N_0(E) = -\frac{2}{\pi}\text{Im}\ln\det\mathbf{M}(E), \tag{6.24}$$

where $\mathbf{M}(E)$ is obtained for the cluster using equation (4.71). Krein's theorem for the cluster, leaving out the bound states, is

$$N(E) - N_0(E) = -2\xi(E) = -\frac{i}{\pi}\ln\det\mathbf{S}(E), \tag{6.25}$$

where $\mathbf{S}(E)$ is the s-matrix for the cluster.

The equations in chapter 2 were derived for a particle scattering from one site, but the same equations apply to a particle scattering from a cluster of atoms. Assuming that **r** is outside a sphere that encompasses the cluster, they would be

$$\psi(E, \mathbf{r}) = \sum_L \psi_L(E, \mathbf{r}) = \frac{1}{2}\left[\sum_L Y_L(\mathbf{r}) h_l^-(\alpha r) + \sum_{L,L'} Y_{L'}(\mathbf{r}) h_{l'}^+(\alpha r) S_{L'L}\right], \tag{6.26}$$

or

$$\psi(E, \mathbf{r}) = \sum_L \psi_L(E, \mathbf{r}) = \frac{1}{2}\left[\sum_L Y_L(\mathbf{r}) j_l(\alpha r) + \sum_{L,L'} Y_{L'}(\mathbf{r}) h_{l'}^+(\alpha r) T_{L'L}\right]. \tag{6.27}$$

These equations contain the same information, simply expressed in a different way. The scattering equations are calculated from the potential function

$$v(\mathbf{r}) = \sum_{i=1}^{N} v_i(\mathbf{r} - \mathbf{R}_i), \tag{6.28}$$

which describes a set of scatterers centered at the positions $\mathbf{R}_i$ in a cluster with a center of mass at some point $\mathbf{R}_0$. Since the $v_i(\mathbf{r} - \mathbf{R}_i)$ are non-overlapping, the particle sees only one of them at a time.

We focused on the t-matrix for the cluster in chapter 3. Under the assumptions outlined in the preceding paragraph, the matrix can be written

$$\mathbf{T} = \mathbf{d}^{-1}\boldsymbol{\tau}\mathbf{d}, \tag{6.29}$$

where $\boldsymbol{\tau}$ is given in equation (4.70) . The elements τ^{ij} are the scattering path operators that go from site i to site j in the cluster. Each operator contains sums over all of the scatterers in the cluster. The elements of **d** are the free particle propagators that connect the terminal site of the scattering path operator to the center of the cluster at $\mathbf{R}_0$. These functions are written out explicitly in a previous reference [8]. Similar arguments apply to the s-matrix.

Since $\det \mathbf{d}^{-1}\mathbf{S}\mathbf{d} = \det \mathbf{d}\mathbf{d}^{-1}\mathbf{S} = \det \mathbf{S}$, we can express the spectral function in terms of the Wigner reaction matrix

$$-2\xi(E) = -\frac{i}{\pi} \ln \det (\mathbf{I} - i\alpha\mathfrak{R})(\mathbf{I} + i\alpha\mathbf{r}\mathfrak{R})^{-1}, \tag{6.30}$$

or

$$\begin{aligned} -2\xi(E) &= -\frac{i}{\pi}[\ln \det (\mathbf{I} + i\alpha\mathfrak{R})^* - \ln \det (\mathbf{I} + i\alpha\mathfrak{R})] \\ &= -\frac{2}{\pi} Im \ln \det (\mathbf{I} + i\alpha\mathfrak{R}), \end{aligned} \tag{6.31}$$

because the reaction matrix is Hermitian. From equation (2.80)

$$-2\xi(E) = -\frac{2}{\pi} Im \ln \det [\mathbf{T}^{-1}\mathfrak{R}] = \frac{2}{\pi} Im \ln \det \mathbf{T}, \tag{6.32}$$

since $\det \mathfrak{R} = 0$. Finally, using equations (6.29) and (6.31) we obtain

$$-2\xi(E) = -\frac{2}{\pi} Im \ln \det \mathbf{M}(E), \tag{6.33}$$

which constitutes a proof that Lloyd's formula is equivalent to Krein's theorem for this special case.

References

[1] Lloyd P 1967 *Proc. Phys. Soc.* **90** 207

[2] Krein M G 1953 *Matem. Sborn.* **33** 597

[3] Birman M L and Krein M G 1962 *Sov. Math.-Dokl.* **3** 740

[4] Wang Y, Stocks G M and Faulkner J S 2014 *J. Phys.: Condens. Matter* **26** 274208

[5] Wang Y, Stocks G M and Faulkner J S 1994 *Phys. Rev.* B **49** 5028

[6] Sutthorp L G and van Wonderen A J 2011 *Opt. Commun.* **284** 2943

[7] Lovatt S C, Gyorffy B L and Guo G Y 1993 *J. Phys.: Condens. Matter* **5** 8005

[8] Faulkner J S 1977 *J. Phys. C. Solid State Phys.* **10** 4661

IOP Publishing

Multiple Scattering Theory
Electronic structure of solids
J S Faulkner, G Malcolm Stocks and Yang Wang

Chapter 7

Toy models

The discussions and equations in the preceding chapters provide a structure for carrying out MST calculations on real solids and using them to explain their physical properties. Clearly, these realistic calculations require massively parallel supercomputers, the faster the better. The reader may find it difficult to get a feel for the process by reading a series of equations. It is for that reason that we introduce toy models that contain the essential physics but are sufficiently simplified that calculations can be done on a personal computer with an application like Mathematica or Matlab®.

7.1 The Kronig–Penney model

Our first toy model is a one-dimensional array of potentials. When the array is periodic in x, this is called the Kronig–Penney (KP) model [1]. We consider a simple version of this model in which the potentials are delta-functions multiplied by a negative coefficient

$$v(x) = -\frac{\hbar^2}{md}\sum_{n=-\infty}^{\infty} P_n\delta(x - na). \tag{7.1}$$

If all of the coefficients of the delta-functions are the same, $P_n = P$, this is the simplest model of a crystalline solid. It was used in the early days of condensed matter physics to demonstrate the concepts of energy bands and band gaps.

7.2 The transfer matrix approach

The Schrödinger equation that can be written in dimensionless units

$$-\frac{d^2\psi}{dx^2} - \frac{2P}{a}\sum_{n=-\infty}^{\infty}\delta\left(x - \left(n + \frac{1}{2}\right)a\right)\psi = E\psi. \tag{7.2}$$

The study of second order differential equations with periodic coefficients dates back many years [2]. The first approach we will discuss is based on the pioneering work of Kramers [3]. The x-axis is broken up into intervals $nd < x \leqslant (n + 1)d$ with a delta-function scatterer at the center of each interval. According to Bloch's theorem [4], the solutions that remain bounded in magnitude satisfy the condition

$$\psi(x + a) = e^{ika}\psi(x), \tag{7.3}$$

so we can focus on the solution in just one interval $0 < x \leqslant d$.

The solution can be written as a combination of two independent solutions

$$\psi(x) = a \cdot u(x) + b \cdot v(x). \tag{7.4}$$

It will turn out to be most convenient to choose the initial conditions

$$\begin{aligned} u(x) &= v'(x) = 1 \\ u'(x) &= v(x) = 0. \end{aligned} \tag{7.5}$$

With these conditions, it follows that

$$a = \psi(0) \qquad b = \psi'(0). \tag{7.6}$$

In region I where $0 < x \leqslant \frac{1}{2}d$, the solutions are

$$\begin{aligned} u_I(x) &= \cos \alpha x \\ v_I(x) &= \frac{1}{\alpha} \sin \alpha x, \end{aligned} \tag{7.7}$$

where $\alpha = \sqrt{E}$.

Integrating equation (7.2) from $\frac{a}{2} - \varepsilon$ to $\frac{a}{2} + \varepsilon$ leads to

$$\begin{aligned} \psi_{II}(0) &= \psi_I(0) \\ \psi_{II}'(0) &= \psi_I'(0) - \frac{2}{a}P\psi(0). \end{aligned} \tag{7.8}$$

These conditions are satisfied if

$$\begin{aligned} u_{II}(x) &= \left(1 + \frac{P}{\alpha a} \sin \alpha a\right) \cos \alpha x - \frac{2P}{\alpha a} \sin^2 \alpha \frac{a}{2} \sin \alpha x \\ v_{II}(x) &= \frac{2P}{\alpha^2 a} \sin^2 \alpha \frac{a}{2} \cos \alpha x + \left(1 + \frac{P}{\alpha a} \sin \alpha a\right) \frac{1}{\alpha} \sin \alpha x. \end{aligned} \tag{7.9}$$

The preceding analysis produces a transfer matrix $\mathbf{T}$ that appears in the relation

$$\begin{pmatrix} \psi_{II}(a) \\ \psi_{II}'(a) \end{pmatrix} = \mathbf{T} \begin{pmatrix} \psi_I(0) \\ \psi_I'(0) \end{pmatrix}. \tag{7.10}$$

The transfer matrix is the Wronskian for this problem. We constructed it so that it is a unit matrix at $x = 0$, so the determinant of $\mathbf{T}$ is real and equal to one. It follows that, when it is diagonalized, the diagonal elements are of magnitude one and are found from the trace of **T**. Following the process outlined above, we obtain

$$\mathbf{T} = \begin{pmatrix} \cos\alpha a - \dfrac{P}{\alpha d}\sin\alpha a & \dfrac{1}{\alpha}\sin\alpha a - \dfrac{2}{\alpha^2 d}P\sin^2\alpha\dfrac{a}{2} \\ -\alpha\sin\alpha a - \dfrac{2}{d}P\cos^2\alpha\dfrac{a}{2} & \cos\alpha a - \dfrac{P}{\alpha d}\sin\alpha a \end{pmatrix}. \tag{7.11}$$

A check on the correctness of this formula is to show algebraically that the determinant of this matrix is one.

Putting all of this together with Bloch's theorem of equation (7.3) leads to the expression for the energy dependence of the Bloch $k(E)$

$$\cos kd = \cos\alpha a - \frac{P}{\alpha a}\sin\alpha a. \tag{7.12}$$

We will return to the transfer matrix method in other contexts.

7.3 The MST approach

Our purpose is to prove Korringa's hypothesis for the simple case of the Kronig–Penney model. We start by assuming an incoming wave from the negative end of the axis, and then show that it can become zero. Using the above diagram, we see that

$$\begin{aligned} \psi_I(x) &= b_n^+ e^{i\alpha x} + c_n e^{-i\alpha x} \\ \psi_{II}(x) &= (b_n^+ + c_n)e^{i\alpha x}. \end{aligned} \tag{7.13}$$

From the continuity of the wave function at $x = 0$ and the discontinuity of the derivative described by equation (7.8) we are led to

$$c_n = ie^{i\vartheta}\sin\vartheta b_n^+, \tag{7.14}$$

with

$$\tan\vartheta = \frac{P}{\alpha a}. \tag{7.15}$$

As in chapter 3 we write the incoming wave as a sum of the outgoing waves from all the other sites,

$$b_0^+ e^{i\alpha x} = \sum_{n=-1}^{-\infty} c_n e^{i\alpha(x-na)} + \sum_{n=1}^{\infty} c_n e^{i\alpha(x+na)} = c_0\left(\sum_{n=-1}^{-\infty} e^{ikna}e^{i\alpha(x-na)} + \sum_{n=1}^{\infty} e^{ikna}e^{i\alpha(x+na)}\right), \tag{7.16}$$

where we have invoked Bloch's theorem

$$c_n = c_0 e^{ikna}. \tag{7.17}$$

Using the standard technique for summing an infinite series of powers, we obtain

$$b_0^+ e^{i\alpha x} = c_0 e^{i\alpha x}(A_- + A_+), \tag{7.18}$$

where

$$A_+ = \frac{e^{ika}e^{iaa}}{1 - e^{ika}e^{iaa}} \quad A_- = \frac{e^{-ika}e^{iaa}}{1 - e^{-ika}e^{iaa}}. \tag{7.19}$$

Combining equation (7.18) with equation (7.14) leads to

$$M(E, k)c_0 = 0, \tag{7.20}$$

with

$$M(E, k) = \left[\frac{\cos ka - \cos \alpha a + \tan \vartheta \sin \alpha a}{(\cos ka - \cos \alpha a)}\right] \tag{7.21}$$

The relation between E and k obtained from the above is exactly the same as that given by the transition matrix method in the preceding section. Korringa's hypothesis is proved because there is no external wave in equation (7.18).

7.4 The Kronig–Penney model of a disordered alloy

A simple model of a substitutional solid solution alloy is obtained by placing the potentials

$$v_A(x) = -\frac{\hbar^2 P_A}{md}\delta(x) \text{ and } v_B(x) = -\frac{\hbar^2 P_B}{md}\delta(x), \tag{7.22}$$

on a line with sites spaced a distance a apart. The randomness is obtained by using a random number generator which provides a set of numbers z such that $0 \leqslant z \leqslant 1$. An alloy with concentrations c_A and $c_B = 1 - c_A$ can be obtained by putting an A atom on a site if $z \leqslant c_A$ and a B atom there otherwise.

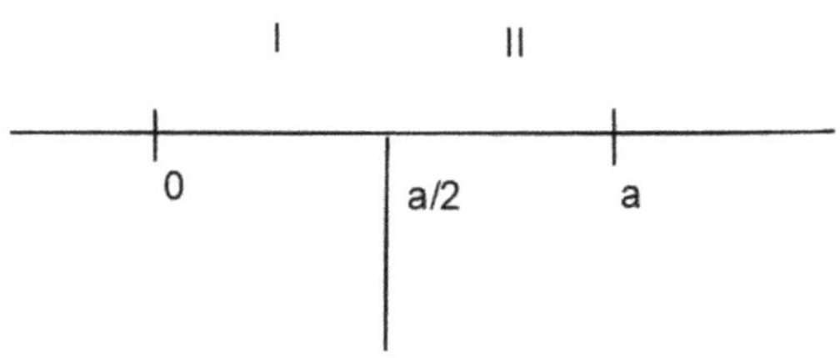

Figure 7.1. A sketch of the cell used in deriving the transfer matrix equations.

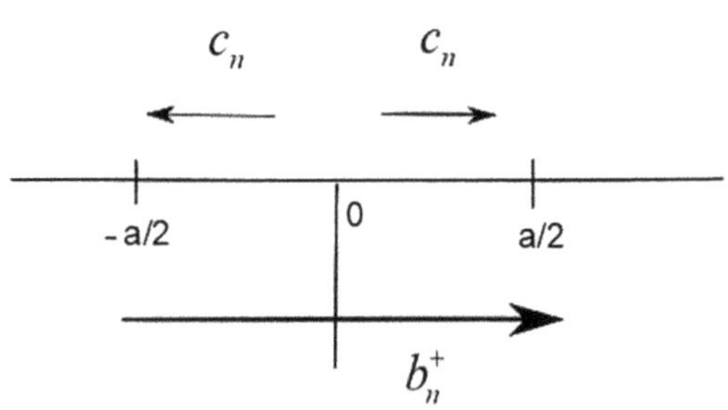

Figure 7.2. A sketch of the cell used in deriving the MST equations.

7.5 The average trace method

The transfer matrix for an alloy model containing N atoms with $N_A = c_A N$ being A atoms and $N_B = c_B N$ B atoms can be written

$$\mathbf{T} = \mathbf{AABBBABAAABB}\ldots \tag{7.23}$$

where $\mathbf{A}$ is obtained from equation (7.11) with $P = P_A$ and $\mathbf{B}$ with $P = P_B$. Since the determinant of $\mathbf{T}$ is one, the integrated density of states for the alloy can be found from the trace of $\mathbf{T}$. James and Ginsbarg [5] devised a useful mathematical algorithm for calculating the IDOS exactly for a given arrangement of A and B atoms.

A method for finding the trace of the average of all possible products of N-r $\mathbf{A}$ matrices and r $\mathbf{B}$ matrices has been derived [6] from Gauss's theorem

$$\langle \mathbf{T} \rangle = \frac{r!(N-r)!}{N!} \oint \frac{(\mathbf{A} + z\mathbf{B})^N}{z^{r+1}} dz. \tag{7.24}$$

This integral is done using the method of steepest descents for the case that N and r become very big. The predictions of this average trace method are shown compared with James and Ginsbarg calculations in figure 7.3; the integrated density of states from the average trace method for a 50% A–B alloy. The δ-function potentials have bound states at minus 1.0 and 0.25 dimensionless units. These calculations were first shown in an extended review entitled 'The Modern Theory of Alloys' [7]. The following discussion of the theory of alloys is supplemented by that publication.

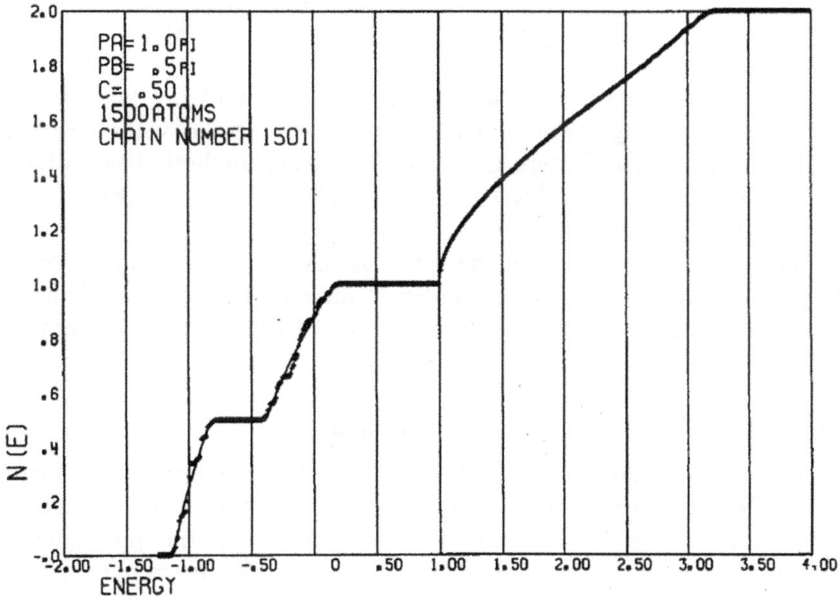

Figure 7.3. The integrated density of states from the average trace method for a 50% A–B alloy. The δ-function potentials have bound states at minus 1.0 and 0.25 dimensionless units. Reprinted from [7]. Copyright (1982), with permission from Elsevier.

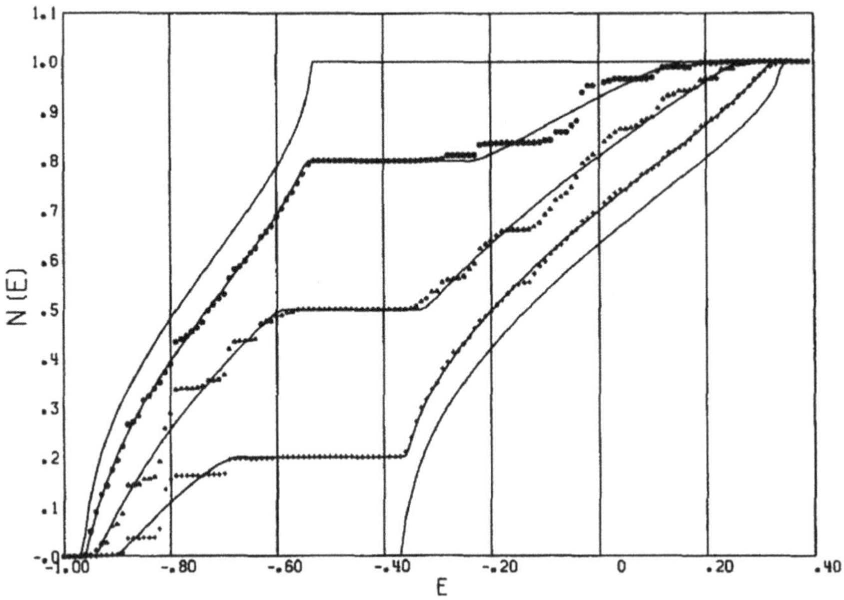

Figure 7.4. The integrated density of states from the average trace method for five examples of an A–B alloy. The δ-function potentials have bound states at minus 0.8 and 0.2 dimensionless units. The concentrations of the B atoms are 0%, 20%, 50%, 80%, and 100%. Reprinted from [7]. Copyright (1982), with permission from Elsevier.

While the average trace method gives excellent overall agreement with the calculated IDOS, there are some fluctuations around its predictions. These appear to be errors in the calculations, but it was shown by Dean [8] that this is not the explanation. He pointed out that, even in a completely random chain of A and B atoms, there will be clusters of atoms like ABAB, ABAABA, …, and these clusters have resonant scattering energies. These clusters will appear with specific probabilities. By studying these clusters, Dean was able to explain both the positions of the fluctuations and their magnitudes. It is important to note that a theory at the level of the average trace method doesn't include this effect.

The most noticeable feature of the IDOS is that, in the low energy region, the bands split into two sub-bands. The number of states in the A or B sub-band is equal to the number of A or B atoms in the crystal. This band-splitting does not occur in the high-energy bands.

7.6 The coherent potential approximation

During the 1960s, a great deal of effort was put into the development of a theory that could describe the electronic structure of realistic three-dimensional systems with no long-range order. Band theory calculations using the muffin-tin approximation had shown that a theory at that level can produce an electronic structure for pure metals that agrees with highly detailed experiments. Techniques like the de Haas van Alphen effect, electron cyclotron resonance, and ultraviolet photoemission spectroscopy give a highly detailed picture of the electronic states in crystalline solids, and first principles band theory calculations agree with these measurements in detail. The figures in the preceding section show that the average trace method predicts an IDOS that agrees very well with exact calculations on an alloy model, but that theory cannot be applied to three-dimensional structures.

The application of ordinary perturbation theory to a model of a binary alloy leads to the virtual crystal approximation. In this approximation, the potential is replaced with a periodic potential with the same symmetry as a pure crystal. The potential that is placed on each lattice site is the average of the A and B potentials

$$v(\mathbf{r}) = c_A v_A(\mathbf{r}) + c_B v_B(\mathbf{r}). \tag{7.25}$$

Without even doing a calculation, it is obvious that this approximation can never reproduce the split bands shown in figure 7.3; the integrated density of states from the average trace method for a 50% A–B alloy. The δ-function potentials have bound states at minus 1.0 and 0.25 dimensionless units.

A somewhat more sophisticated theory was proposed by Korringa [9] and later by Beeby [10]. This average t-matrix approximation, ATA, is based on the multiple scattering band theory described in chapter 4. Again, the disordered system is replaced by an ordered system. The t-matrix that describes the scattering from each site is the average of the A and B t-matrices

$$t(E) = c_A t_A(E) + c_B t_B(E). \tag{7.26}$$

Comparisons with one-dimensional model calculations show that the ATA is actually worse than the virtual crystal method because it introduces structures into the calculated density of states that have no meaning.

These unsuccessful theories made it clear that ordinary perturbation theory was not sufficient for the purpose of developing a theory of electronic states in a disordered alloy. The reason is that the character of the desired solutions is not the same as that for the initial states and, in fact, the character of the desired solutions is not even known. Luckily, this theoretical challenge had arisen before, and the answer had been found by Feynman and others. The answer had to come from infinite order perturbation theory, and the contributions to the solution would be analyzed using a version of Feynman diagrams.

Klauder [11] carried out a diagrammatic analysis that showed that the contribution that was missing in the ATA is a form of self-consistency in the treatment of the propagation of the electron. The electron needed to know that it was in an alloy between scattering events and not just at the scattering event. In the language of diagrams, this is known as internal propagator modification.

Klauder derived his DOS formula for a one-dimensional model of a liquid metal. The potential function is

$$v(x) = -\frac{\hbar^2}{md} \sum_{n=-\infty}^{\infty} P\delta(x - x_n), \tag{7.27}$$

where identical δ-function potentials are placed at a random set of sites. The sites, x_n, are distributed along the x-axis according to the Poisson distribution. The average distance between sites is d. The node counting method has been adapted to give exact calculations for this model [12]. When the average trace method is similarly adapted, it can be shown to give a formula that is algebraically identical to the Klauder formula [13].

The culmination of the theoretical progress that was made in the early 1960s is a theory known as the coherent potential approximation (CPA). In reference [1], chapter 1, Soven describes the CPA verbally in the following way. An alloy that has potentials v_A and v_B distributed randomly on the A and B sites is replaced by a periodic system with an effective potential v_{cpa} on every site. The effective potential is chosen such that, on average, the scattering from an impurity A or B atom embedded in a crystal with v_{cpa} on all other sites will be zero. In the language of MST, this description leads to equation (5.11). Soven did some one-dimensional calculations to illustrate its usefulness, and it was later realized that his result was equivalent to the average trace method and Klauder's formula. Perhaps the most convincing arguments come from the formulation of the problem of calculating the Green's function of an alloy in terms of infinite order perturbation theory, as pointed out by Klauder. Diagrams are used to analyze the contributions to the Green's functions in a manner analogous to field theory, quantum electrodynamics, or many-body theory. It has been argued that the CPA contains the maximum number of classes of diagrams that can be included and still retain a Green's function that is properly analytic. The criterion that is used to specify the diagrams included in the CPA is called the single-site approximation. It has been shown by Mills and Ratanavararaksa [14] and also Mookerjee [15] that it is possible to go beyond the single-site approximation and retain an analytic Green's function, but it is difficult.

The most elaborate application of MST to the problem of one-dimensional alloys is in a paper by Butler [16]. He did calculations on a complicated model that was designed to emulate a d-band transition metal. Butler's potential leads to two-by-two matrix equations, but, when a δ-function potential is used, they become one-dimensional. The t-matrix for this case is

$$t_0 = -\alpha e^{i\vartheta} \sin\vartheta, \tag{7.28}$$

with $\tan\vartheta$ given by equation (7.15). The Green's function is

$$G_{00}(E, k) = \frac{1}{i\alpha}\left[-1 + \frac{i \sin\varphi}{\cos\vartheta - \cos\varphi}\right]. \tag{7.29}$$

For a model with only one type of potential, the Korringa matrix in equation (3.42) becomes exactly the same as the one in equation (7.21) that was obtained from the MST approach.

The scattering path matrix for a model that has the effective CPA scattering strength P_c on every site is

$$\tau_c^{00} = -\frac{P_c}{a}\left(1 - i\frac{P_c}{\varphi}\frac{\sin\varphi}{\sqrt{1-\chi^2}}\right). \tag{7.30}$$

The matrices for the case where there is an A or B atom on one site are

$$\begin{aligned}\tau_A^{00} &= \left[1 + \tau_c^{00}(x_A - x_c)\right]^{-1}\tau_c^{00}\\ \tau_B^{00} &= \left[1 + \tau_c^{00}(x_B - x_c)\right]^{-1}\tau_c^{00},\end{aligned} \tag{7.31}$$

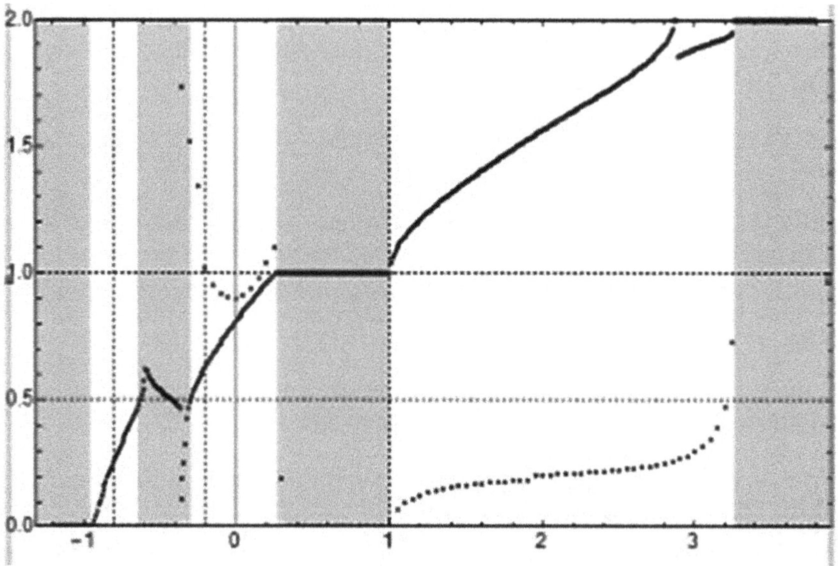

Figure 7.5. The integrated density of states from the average trace method and the density of states from the CPA for a 50% A–B alloy. The δ-function potentials have bound states at minus 0.8 and 0.2 dimensionless units. The shaded areas show the energies for which the number of states is zero.

where

$$x_A = \frac{a}{P_A} \quad x_B = \frac{a}{P_B} \quad x_c = \frac{a}{P_c}. \tag{7.32}$$

The CPA equation (5.11) becomes

$$x_c = c_A x_A + c_B x_B + \tau_c^{00}(x_A - x_c)(x_B - x_c). \tag{7.33}$$

Obviously, this equation has to be solved by iteration. The results of the CPA calculation are compared with those of an average trace calculation in figure 7.5.

We show the 'raw' calculations that contain discontinuities and computational anomalies. In the average trace method, these arise from the fact that the saddle points that appear in the steepest descents method are the complex solutions of a cubic equation. They can jump from one Riemann sheet to another in unpredictable ways, and this magnifies any lack of convergence in the calculation. The main problem with the CPA calculations is the difficulty in converging an iterative solution of equation (7.33).

7.7 Lloyd's formula for the Kronig–Penney model

From equations (6.4) and (7.21), Lloyd's formula for this model is

$$N(E) - N_0(E) = -\frac{2}{\pi}\mathrm{Im}\ln\left[\frac{\cos ka - \cos \alpha a + \tan \vartheta \sin \alpha a}{(\cos ka - \cos \alpha a)}\right]. \tag{7.34}$$

The numerator and denominator of the argument of the logarithm are both real, so the only imaginary quantities are the odd multiples of π that appear when the functions are negative. It follows that

$$N(E) = \frac{4k(E)}{\pi}, \tag{7.35}$$

where the $k(E)$ are the boundaries of the range within which $\cos ka \leqslant \cos \alpha a + \tan \vartheta \sin \alpha a$. Similarly,

$$N_0(E) = \frac{4k_0(E)}{\pi}, \tag{7.36}$$

where the $k_0(E)$ are the limits of the range where $\cos ka \leqslant \cos \alpha a$. These formulae for the integrated densities of states are clearly correct.

7.8 The spherical square well

The relations between Krein's theorem, the Friedel sum, Levinson's theorem, and integrals over the Green's function for a single scatterer embedded in a vacuum were discussed in the previous chapter. They were illustrated with calculations on non-spherical potentials for real elements. From a purely mathematical point of view, the easiest way to illustrate the algebra is to do some calculations on a spherical square well potential [17]. The potential is $v(\mathbf{r}) = -\Delta$ for $|\mathbf{r}| < R$ and $v(\mathbf{r}) = 0$ otherwise. The advantage of this model is that all integrals that appear in the calculations can be done analytically, and any hypothesis can be checked numerically in a matter of minutes using a personal computer with an application like Mathematica or Matlab®.

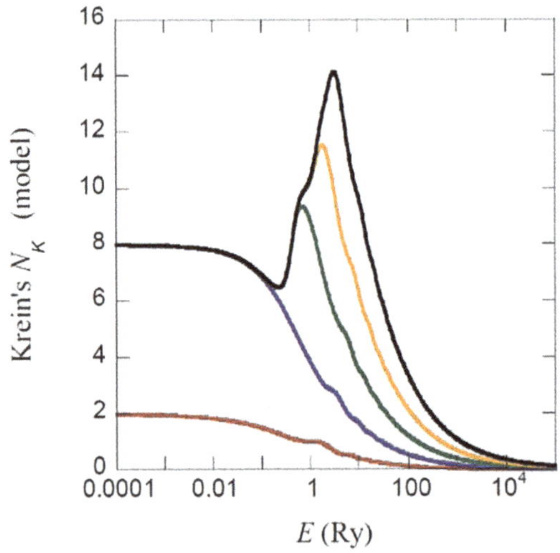

Figure 7.6. The curves show the Krein IDOS $N_K(E)$ for $\Delta = 1.4$ (black). The lowest (red) curve was calculated for $l_{max} = 0$. The blue one is for $l_{max} = 1$, while the green, orange, and black curves correspond to $l_{max} = 2$, $l_{max} = 3$, and $l_{max} = 4$. Reproduced from [3]. © 2014 IOP Publishing Ltd. CC BY 3.0.

The radius R of the wells we use is chosen to be 3 because that is approximately the radius of atomic spheres in a metal in units of Bohr radii. Because of this, the dimensionless energies in the model calculations have the same scale as the Rydbergs used in a real solid. The square well potentials have a bound state for $l = 0$ when the magnitude of Δ becomes greater than 0.274155678. When Δ becomes greater than 1.096622711, it has a bound state with $l = 1$, while the $l = 2$ bound state appears when Δ is greater than 2.243414825.

Writing the elements of the S-matrix as in equation (6.15), the Krein IDOS for a given l_{max} has the form

$$N_K(E) = \frac{2}{\pi}\sum_{l=0}^{l_{max}}(2l + 1)\delta_l(E). \tag{7.37}$$

This function is plotted in figure 7.6 for energies ranging from $E_{min} = 0.0001$ and $E_{max} = 100\ 000$. The model has $\Delta = 1.4$ so the well binds 8 electrons, 2 for $l = 0$ and 6 for $l = 1$. The curves shown correspond to $0 \leqslant l_{max} \leqslant 4$. Those for $l_{max} \geqslant 1$ illustrate Levinson's theorem in the form of equation (6.13). The origin of the structure in $N_K(E)$ at $E \geqslant 0.85$ is scattering resonances for $l \geqslant 2$. As the depth of the well increases, these resonances appear at lower energies until, at some Δ, they become bound states.

The function

$$\mathrm{n}_{in}(E) = -\frac{2}{\pi}\int_0^R \mathrm{Im}[G(E, r, r) - G_0(E, r, r)]r^2 dr, \tag{7.38}$$

is called a partial density of states because the integral is only over the sphere. This is equivalent to a partial trace. It follows that

$$N_{in}(E) = \int_{E_B}^{E} n_{in}(E')dE', \tag{7.39}$$

is a partial IDOS. The free particle Green's function is removed in equation (7.38) so that the result can be compared with the Krein IDOS.

In figure 7.7, $N_K(E)$ and $N_{in}(E)$ are plotted for three values of Δ. The bound states have the degeneracy $d_l = 2(2l + 1)$.

The potential with $\Delta = 1.4$ models a system with 8 s and p states, but is not close to binding d states. The potential with $\Delta = 2.1$ is close to binding d states, and the one for $\Delta = 2.4$ has just bound the 10 d states. The abrupt step in $N_K(E)$ for $\Delta = 2.1$ is the precursor of the change in $\xi(0)$ expected from Levinson's theorem when Δ is greater than 2.243414825. A glance at the periodic table shows that molybdenum would be expected to have an $N_K(E)$ like our model for $\Delta = 1.4$, while copper should be similar to the one for $\Delta = 2.1$ and aluminum is similar to the $\Delta = 2.4$ case. Calculations on real materials are shown in the previous chapter, and these hypotheses are shown to be true.

It is clear from figure 7.8 that the partial IDOS $N_{in}(E)$ is not very close to $N_K(E)$. The trace operation discussed in connection with equation (7.38) can be completed by adding the integral

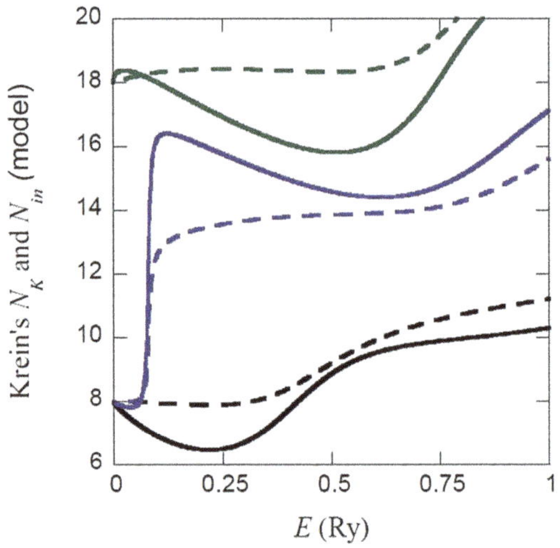

Figure 7.7. The solid lines show the Krein IDOS $N_K(E)$ for $\Delta = 1.4$ (black), $\Delta = 2.1$ (blue), and $\Delta = 2.4$ (green). The dotted lines show the partial IDOS $N_{\text{in}}(E)$ for the same $\Delta's$ calculated by integrating the Green's function within the sphere with radius R. Since the contribution from the free-electron Green's function does not appear in the $N_K(E)$, it is removed from the GF-IDOS. Reproduced from [3]. © 2014 IOP Publishing Ltd. CC BY 3.0.

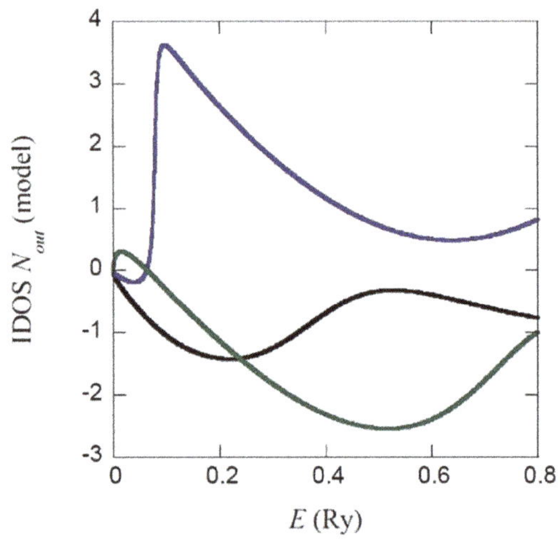

Figure 7.8. The solid lines show the partial IDOS $N_{\text{out}}(E)$ for $\Delta = 1.4$ (red), $\Delta = 2.1$ (blue), and $\Delta = 2.4$ (green). Reproduced from [3]. © 2014 IOP Publishing Ltd. CC BY 3.0.

$$n_{\text{out}}(E) = -\frac{2}{\pi}\int_{\Omega_\infty - \Omega} \text{Im}[G(E, \mathbf{r}, \mathbf{r}) - G_0(E, \mathbf{r}, \mathbf{r})]d\mathbf{r}, \tag{7.40}$$

where the volume $\Omega_\infty - \Omega$ is all of the space outside the domain in which the potential is nonzero. The corresponding partial IDOS is then

$$N_{\text{out}}(E) = \int_{E_B}^{E} n_{\text{out}}(E')dE'. \tag{7.41}$$

From the algebra in chapter 4, the Green's function for $\mathbf{r} = \mathbf{r}'$ takes the form

$$G(E, \mathbf{r}, \mathbf{r}) = G_0(E, \mathbf{r}, \mathbf{r}) - \alpha^2 \sum_{L,L'} Y_L(\mathbf{r}) h_l(\alpha r) t_{LL'}(E) h_{l'}(\alpha r) Y^*_{L'}(\mathbf{r}). \tag{7.42}$$

In this equation, $h_l(\alpha r)$ is the Hankel function that describes outgoing waves. The contribution to the DOS from the outside integral is therefore

$$n_{\text{out}}(E) = \frac{2\alpha^2}{\pi} \text{Im} \sum_{L,L'} \int_{\Omega_\infty - \Omega} \left[Y_L(\mathbf{r}) h_l(\alpha r) t_{LL'}(E) h_{l'}(\alpha r) Y^*_{L'}(\mathbf{r}) \right] d\mathbf{r}. \tag{7.43}$$

For a spherical potential, this becomes

$$n_{\text{out}}(E) = \frac{2\alpha^2}{\pi} \text{Im} \sum_l (2l + 1) t_l(E) \int_R^\infty h_l(\alpha r)^2 r^2 dr, \tag{7.44}$$

where

$$t_l(E) = -\frac{1}{\alpha} e^{-i\eta_l} \sin \eta_l. \tag{7.45}$$

The advantage in writing $n_{\text{out}}(E)$ in this form is that the definite integral in equation (7.44) has been worked out and published, as shown in connection with equation (6.21).

The partial IDOS $N_{\text{out}}(E)$ obtained from these equations for the three values of Δ are shown in figure 7.8; a visual comparison of figure 7.7. The solid lines show the Krein IDOS $N_K(E)$ for $\Delta = 1.4$ (black), $\Delta = 2.1$ (blue), and $\Delta = 2.4$ (green). The dotted lines show the partial IDOS $N_{\text{in}}(E)$ for the same calculated Δ_s' by integrating the Green's function within the sphere with radius R. Since the contribution from the free-electron Green's function does not appear in the $N_K(E)$, it is removed from the GF-IDOS. confirms that the Krein IDOS is the same as the sum of the two partial IDOSs

$$N_K(E) = N_{\text{in}}(E) + N_{\text{out}}(E), \tag{7.46}$$

as it should be.

The discerning eye will notice that the total IDOS shown in figure 7.7 has a typical behavior at small energies. The reason for this can be seen from equation (7.37). It is well known that the phase shifts for well-behaved spherical potentials approach zero like

$$\lim_{E \to 0} \delta_l(E) = c_l E^{(l+1/2)}. \tag{7.47}$$

The Krein IDOS will be dominated by the $l = 0$ phase shift, which has the limit

$$\lim_{E \to 0} \delta_0 \to c_0 E^{1/2}. \tag{7.48}$$

The Krein DOS, obtained by numerical differentiation, must then approach infinity like

$$\lim_{E\to 0} n_K(E) \to \frac{c_0}{\pi} E^{-1/2}. \tag{7.49}$$

The partial DOS $n_{\text{in}}(E)$ approaches zero smoothly, so it follows that $n_{\text{out}}(E) = n_K(E)$ for small energies. These observations hold for the calculations on real potentials in the preceding chapter, but they can be seen analytically for this simple model.

References

[1] de R, Kronig L and Penney W G 1931 *Proc. R. Soc. (London)* A **130** 499
[2] Floquet G 1883 *Ann. de l'École Norm. Supér.* **12** 47–88
[3] Kramers H A 1935 *Physica* **2** 483
[4] Felix B 1928 *Z. Phys.* **52** 555–600
[5] James H M and Ginsbarg A S 1953 *J. Phys. Chem.* **57** 849
[6] Faulkner J S and Korringa J 1964 *Phys. Rev.* **122** 390
[7] Faulkner J S 1982 *Progress in Materials Science* Vol. 27 ed J W Christian, P Haasen and T B Massalski (Oxford: Pergamon)
[8] Dean P 1960 *Proc. R. Soc. (London)* **A254** 507
[9] Korringa J 1958 *J. Phys. Chem. Solids* **7** 252
[10] Beeby J L 1964 *Phys. Rev.* **135** A130
[11] Klauder J R 1961 *Ann. Phys. (New York)* **14** 43
[12] Lax M and Philips J C 1958 *Phys. Rev.* **110** 41
[13] Faulkner J S 1964 *Phys. Rev.* **135** A124
[14] Mills R and Ratanavararaksa P 1978 *Phys. Rev.* B **18** 5291
[15] Mookerjee A 1973 *J. Phys.* C **6** 1340
[16] Butler W H 1976 *Phys. Rev.* B **14** 468
[17] Wang Y, Stocks G M and Faulkner J S 2014 *J. Phys.: Condens. Matter* **26** 274208

IOP Publishing

Chapter 8

Relativistic full potential MST calculations

From the earliest days of quantum mechanics, it was understood that the Schrödinger equation is incomplete because it is not consistent with the theory of relativity. Sommerfeld [1] used that theory to explain the fine structure of the spectrum of hydrogen. It was understood that some of the effects of relativity are to cause an orbital contraction of the *s* and *p* atomic shells, which then leads to an expansion of the *d* and *f* shells. The appearance of a magnetic anisotropy energy in iron, for example, can be explained only by relativistic calculations.

8.1 The Dirac equation

The derivation of new equations that are consistent with relativity starts from the equation for the energy

$$W^2 = p^2c^2 + m^2c^4. \tag{8.1}$$

Making the substitution that proved successful for the Schrödinger case

$$p_x = -i\hbar\frac{\partial}{\partial x}, \quad W = i\hbar\frac{\partial}{\partial t} \tag{8.2}$$

leads to the Klein–Gordon equation

$$\nabla^2\phi - \frac{1}{c^2}\frac{\partial\phi}{\partial t^2} = \frac{m^2c^2}{\hbar^2}\phi. \tag{8.3}$$

The second time derivative leads to problems in quantum mechanics if we want to keep the interpretation of the absolute square of the wave as the time dependent probability for the location of a particle. A generalization of the probability density can be made, but it is not positive definite. This is one of the things that led Dirac to derive another equation that did not have this fault.

The square root of equation (8.1) is

doi:10.1088/2053-2563/aae7d8ch8

$$W = c\boldsymbol{\alpha} \cdot \mathbf{p} + \beta mc^2, \tag{8.4}$$

if the alphas and beta satisfy the anticommutation rules $\{\alpha_i, \alpha_j\} = 2\delta_{ij}$, $\{\alpha_i, \beta\} = 0$, and $\beta^2 = 1$. It has been known since the days of Hamilton that these conditions are satisfied by the four-by-four matrices

$$\alpha_i = \begin{pmatrix} 0 & \sigma_i \\ \sigma_i & 0 \end{pmatrix} \qquad \beta = \begin{pmatrix} \mathbf{1} & 0 \\ 0 & -\mathbf{1} \end{pmatrix}, \tag{8.5}$$

where σ_i is a two-by-two Pauli spin matrix and $\mathbf{1}$ is a two-by-two unit matrix.

Starting from equation (8.4), the Dirac equation for an electron in an electromagnetic field is written

$$(W - e\Phi)\psi = c\boldsymbol{\alpha} \cdot \left(\mathbf{p} - \frac{e}{c}\mathbf{A}\right)\psi + \beta mc^2\psi, \tag{8.6}$$

where ψ is a four-dimensional vector that is normally broken into two parts

$$\psi = \begin{pmatrix} u_b \\ u_s \end{pmatrix}. \tag{8.7}$$

The two dimensional vector u_b is called the large component, and u_s is the small component. Equation (8.6) can be broken into two parts

$$\begin{aligned} (W - e\Phi - mc^2)u_b &= c\boldsymbol{\sigma} \cdot \left(\mathbf{p} - \frac{e}{c}\mathbf{A}\right)u_s \\ (W - e\Phi + mc^2)u_s &= c\boldsymbol{\sigma} \cdot \left(\mathbf{p} - \frac{e}{c}\mathbf{A}\right)u_b. \end{aligned} \tag{8.8}$$

The small component can be eliminated from these equations to obtain

$$\begin{aligned} Eu_b = &\frac{1}{2m}\left(\mathbf{p} - \frac{e}{c}\mathbf{A}\right)^2 u_b + Vu_b - \frac{e\hbar}{2mc}\boldsymbol{\sigma} \cdot \mathbf{B}u_b \\ &- \frac{(\varepsilon - V)^2}{2mc^2}u_b + \frac{\hbar^2}{2m}(E - V + 2mc^2)^{-1} \\ &[\nabla V \cdot \nabla u_b + i\boldsymbol{\sigma} \cdot \nabla V \times \nabla u_b] \\ &- i\frac{e\hbar}{2mc}(E - V + 2mc^2)^{-1} \\ &[\nabla V \cdot \mathbf{A} + i\boldsymbol{\sigma} \cdot \nabla V \times \mathbf{A}]u_b, \end{aligned} \tag{8.9}$$

where $E = W - mc^2$, and we have replaced $e\Phi(\mathbf{r})$ with $V(\mathbf{r})$. The top row of this expression is a self-contained equation called the Pauli equation. The first term in the second row is known as the mass-velocity correction. The next term in this row is called the Darwin term, and the cross-product term describes the spin–orbit coupling. The two terms in the last row are small, and do not appear if $\mathbf{A}$ is zero.

If there is no magnetic field in the system we are considering, the Pauli equation becomes the Schrödinger equation, and the MST described in the preceding chapters applies without change. An approximation that is frequently used to deal with equation (8.9) with no magnetic field is to include everything except for the spin–orbit term. This scalar relativistic approximation (SRA) again leads to a Schrödinger equation except that the potential is replaced by

$$U(\mathbf{r}) = V(\mathbf{r}) - \frac{(E - V(\mathbf{r}))^2}{2mc^2} + \frac{\hbar^2}{2m}\left(E - V(\mathbf{r}) + 2mc^2\right)^{-1}\nabla V(\mathbf{r}) \cdot \nabla. \qquad (8.10)$$

The mass-velocity correction makes the potential look somewhat more complicated, but it introduces no fundamental difficulties. The gradient operator in the Darwin term is different from anything we have seen before. Formulae for the sine and cosine matrices that have been modified to deal with this term have been derived [2].

The idea of the SRA is to leave out the complicated spin–orbit term in the MST calculation and put it in later using perturbation theory. This works quite well for many purposes, but it would obviously be better to use the Dirac equation in full potential MST calculations without approximations. The way to do that has been shown in a recent publication.

A detailed analysis of the Dirac equation can be found in the book by Dr P Strange [3]. An abbreviated discussion of MST based on the Dirac equation also appears in the book, as well as a description of the magnetic anisotropy energy in magnetic metals.

8.2 Relativistic Green's function

Numerous publications on relativistic band theory can be found in the literature. Some of these are based on MST [4]. As can be surmised from the preceding chapters, modern MST condensed matter calculations have moved away from the KKR band theory method to a more general use of Green's functions. This is particularly true for full potential calculations (see reference [28] in chapter 1).

A derivation of the full potential MST Green's function based on the Dirac equation (RFP-MST) has been described in some recent publications [5–7]. It is based on the fact that the principles of scattering theory are the same whether the wave function comes from the Dirac equation or the Schrödinger equation. The authors start by defining Dirac analogs to the functions that are used in chapter 4. They then define sine and cosine matrices, incoming and outgoing wave functions, and Green's functions that look formally very much like the ones described in chapter 4.

The development starts by rewriting the Dirac equation, equation (8.6), in dimensionless units

$$\left[-ic\boldsymbol{\alpha} \cdot \nabla + V(\mathbf{r}) + \frac{1}{2}\beta c^2\right]\psi = W\psi, \qquad (8.11)$$

with $\hbar = 1$, length in Bohr radii, and energy in Rydbergs. In these units, the velocity of light is twice the inverse of the fine structure constant $c = 274.072$. The solutions of this equation with $V(\mathbf{r}) = 0$ that are analogous to those in chapter 4 are

$$J_\Lambda(E, \mathbf{r}) = \left(\frac{W}{c^2} + \frac{1}{2}\right)^{1/2} \begin{pmatrix} j_l(pr)\chi_\Lambda(\hat{\mathbf{r}}) \\ \dfrac{ipcS_\kappa}{W + \dfrac{c^2}{2}} j_{\bar{l}}(pr)\chi_{\bar{\Lambda}}(\hat{\mathbf{r}}) \end{pmatrix}, \tag{8.12}$$

and

$$N_\Lambda(E, \mathbf{r}) = \left(\frac{W}{c^2} + \frac{1}{2}\right)^{1/2} \begin{pmatrix} n_l(pr)\chi_\Lambda(\hat{\mathbf{r}}) \\ \dfrac{ipcS_\kappa}{W + \dfrac{c^2}{2}} n_{\bar{l}}(pr)\chi_{\bar{\Lambda}}(\hat{\mathbf{r}}) \end{pmatrix}. \tag{8.13}$$

In these equations, l is the ordinary angular momentum index, $\Lambda = (\kappa, \mu)$, $\bar{\Lambda} = (-\kappa, \mu)$, $S_\kappa = \kappa / |\kappa|$, and

$$\bar{l} = \begin{Bmatrix} l + 1 \text{ if } \kappa < 0 \\ l - 1 \text{ if } \kappa > 0 \end{Bmatrix}. \tag{8.14}$$

The spin-angular functions are

$$\chi_\Lambda(\hat{\mathbf{r}}) = \sum\nolimits_{m_s=-1/2}^{1/2} C\left(l, j, \frac{1}{2} | \mu - m_s, m_s\right) Y_{l,\, \mu - m_s}(\hat{\mathbf{r}})\phi_{m_s}, \tag{8.15}$$

and

$$\chi_{\bar{\Lambda}}(\hat{\mathbf{r}}) = \sum\nolimits_{m_s=-1/2}^{1/2} C\left(\bar{l}, j, \frac{1}{2} | \mu - m_s, m_s\right) Y_{\bar{l},\, \mu - m_s}(\hat{\mathbf{r}})\phi_{m_s}, \tag{8.16}$$

where the $C\left(l, j, \frac{1}{2} | \mu - m_s, m_s\right)$ are Clebsch–Gordon coefficients. We see that $J_\Lambda(E, \mathbf{r})$ and $N_\Lambda(E, \mathbf{r})$ are 4×1 vectors because the ϕ_{m_s} are 2×1 Pauli vectors

$$\phi_{1/2} = \begin{pmatrix} 1 \\ 0 \end{pmatrix}, \qquad \phi_{-1/2} = \begin{pmatrix} 0 \\ 1 \end{pmatrix}. \tag{8.17}$$

This morass of subscripts has always made the Dirac equation appear foreboding. In modern times, the equations are easier to deal with because computers take on a lot of the load. Once the notation is developed, the solution of the Dirac equation that is used in the RFP-MST appears very much like the one in equation (2.51)

$$\psi_\Lambda(E, \mathbf{r}) = \sum [N_{\Lambda'}(E, \mathbf{r}) S_{\Lambda'\Lambda}(E, r) - J_{\Lambda'}(E, \mathbf{r}) C_{\Lambda'\Lambda}(E, r)], \tag{8.18}$$

with

$$C_{\Lambda'\Lambda}(E, r) = p \int_{|\mathbf{r}'| \leqslant r} N_{\Lambda'}^{\dagger}(E, \mathbf{r}') V(\mathbf{r}') \psi_\Lambda(E, \mathbf{r}') d\mathbf{r}' - \delta_{\Lambda'\Lambda}, \tag{8.19}$$

and

$$S_{\Lambda'\Lambda}(E, r) = p\int_{|\mathbf{r}'|\leqslant r} N^{\dagger}_{\Lambda'}(E, \mathbf{r}')V(\mathbf{r}')\psi_{\Lambda}(E, \mathbf{r}')d\mathbf{r}', \tag{8.20}$$

where $p = \sqrt{E}$. As pointed out, equation (2.51) is the result of an extension of Calogero's method to include non-spherical potentials, and equation (8.18) represents an extension of that method to include nonrelativistic non-spherical potentials,

The single-site Green's function for this case is written

$$G(E, \mathbf{r}, \mathbf{r}') = \sum_{\Lambda,\Lambda'} Z_{\Lambda}(E, \mathbf{r})t_{\Lambda\Lambda'}(E)Z^{\dagger}_{\Lambda'}(E, \mathbf{r}') - \sum_{\Lambda} Z_{\Lambda}(E, \mathbf{r})J^{\dagger}_{\Lambda}(E, \mathbf{r}'), \tag{8.21}$$

where

$$Z_{\Lambda}(E, \mathbf{r}) = p\sum_{\Lambda'}\psi_{\Lambda'}(E, \mathbf{r})S^{-1}_{\Lambda'\Lambda}(E). \tag{8.22}$$

The t-matrix is analogous to the one in equation (2.50),

$$t_{\Lambda\Lambda'}(E) = -\frac{1}{p}\sum_{\Lambda''} S_{\Lambda\Lambda''}(E)[\mathbf{C} - i\mathbf{S}]^{-1}_{\Lambda''\Lambda'}, \tag{8.23}$$

and $J_{\Lambda}(E, \mathbf{r})$ is calculated by an inward integration of the Dirac equation just as the function $J_L(E, \mathbf{r})$ was obtained using the Schrödinger equation.

This Green's function is used to calculate the charge density and density of states just like the simpler functions derived in chapter 4. Four methods were outlined in that chapter for doing the energy integrals that are required for calculating the charge density or the total charge from $G(E, \mathbf{r}, \mathbf{r}')$. In the RFP-MST, the fourth method is used. The poles of $[\mathbf{C} - i\mathbf{S}]^{-1}$ are found and used to improve the energy integrals.

There is a lot of new technology in the RFP-MST. It is tested in reference [5] by comparing DOS and IDOS calculations from the Green's function in equation (8.21) with those from the application of Krein's theorem using the S-matrix. As described in chapter 6 for the nonrelativistic case, the agreement is precise. This demonstrates that there are no errors in the formulation.

The primary difference between the RFP-MST described in references [5] and [6] and previous efforts is, as discussed in chapter 4, that the wave functions in the vicinity of the nucleus are treated correctly, which guarantees that all the eigenstates in the MST are properly orthogonal to the eigenstates with lower energies.

8.3 Some examples

There are many examples of RFP-MST calculations in references [5] and [6], but the one that best illustrates the effects of a full potential as over against a spherical potential, and a relativistic as over against a nonrelativistic calculation is the single-site DOS of silver. This is shown in figure 8.1.

The very sharp peak at approximately 0.06 Ry is due to states that have been split off primarily by the effect of spin–orbit coupling. The double maxima in the distribution of higher energy states from both the relativistic and nonrelativistic

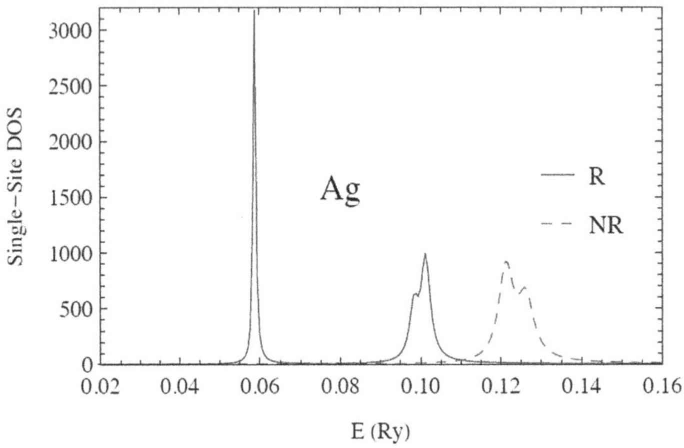

Figure 8.1. Comparison of the relativistic and nonrelativistic single-site DOS of silver. The blue solid line is calculated from the RFP-MST method. The red dotted line is calculated using the nonrelativistic full potential Green's function method. Reproduced from [6]. © 2016 IOP Publishing Ltd.

calculations appears only in full potential calculations. The difference between the results of full potential calculations and those in which the potential has been approximated by a simpler one is larger for the relativistic case.

The importance of the development of reliable techniques for full potential MST cannot be overestimated. Most MST calculations to date have been carried out on metals and metallic alloys. In the future, a greater understanding of the properties of such systems as covalently bonded solids can be studied.

References

[1] Sommerfeld A 1916 Ann. Phys., Lpz. **51** 125

[2] Faulkner J S 1994 *Solid State Commun.* **90** 791

[3] Strange P 1998 *Relativistic Quantum Mechanics* (Cambridge: Cambridge University Press)

[4] Xindong Wang X-G, Zhang W, Butler H, Stocks G M and Harmon B N 1992 *Phys. Rev.* B **46** 9352

[5] Liu X, Wang Y, Eisenbach M and Stocks G M 2018 *Comput. Phys. Commun.* **224** 265

[6] Liu X, Wang Y, Eisenbach M and Stocks G M 2016 *J. Phys.: Condens. Matter* **28** 355501

[7] Liu X 2017 *PhD dissertation* Carnegie Mellon University

IOP Publishing

Multiple Scattering Theory
Electronic structure of solids
J S Faulkner, G Malcolm Stocks and Yang Wang

Chapter 9

Applications of MST

We have already mentioned some applications of MST in the process of explaining the method and deriving the basic equations. The focus is on calculations that MST is uniquely suited for. The precision of MST equations is outstanding, especially in comparison with linearized methods that have been simplified for the purpose of making them run faster on intermediate performance computers. Unlike methods based on the tight-binding method, bonding states in the conduction or valence band are guaranteed to be orthogonal to lower level core states. Constant energy searches can be carried out in the Brillouin zone. In chapter 3, the way that aspect of MST can be used to obtain extremely high precision calculations of the density of states in the region of the Fermi energy was covered. MST is well suited for the calculation of Green's functions in condensed matter. This is useful for a wide range of calculations, including a formalism for studying impurities embedded into an otherwise periodic crystal. It is the basis for the alloy theory called the KKR-CPA, as pointed out in chapter 5. The locally self-consistent MST method, which can be used to treat systems with and without long-range order, relies on the Green's function formulation. This was also discussed in chapter 5. Calculations on phenomena that we now know are caused by spin–orbit coupling rely on the relativistic version of MST.

To understand the importance of some of the most impressive applications of MST to treat phenomena in condensed matter physics and materials science, it is necessary to recall the historical context. In the 1960s, the development of new materials for modern applications was the work of physical metallurgists and materials chemists [1]. The only theoretical tool was thermodynamics. Later x-ray diffraction and electron microscopy were added to the toolbox. It was understood at the time that the glue that holds materials together is a cloud of conduction or valence electrons, but it appeared unlikely that electronic structure calculations would be of practical importance in explaining the phenomena that are of interest to materials scientists. For this reason, a formal course in quantum mechanics was rarely included in the curriculum for metallurgy and chemistry students.

doi:10.1088/2053-2563/aae7d8ch9

At the same time, progress was being made by condensed matter theorists. Much to the surprise of many of the biggest names in physics, the one-electron theory and early versions of density functional theory were making it possible to do detailed calculations of the Fermi surfaces in metals that agreed with extremely accurate measurements [2]. That theory also provided an understanding of the crystal structure and even the lattice constants of crystals. Studies of the electronic structure of semiconductor materials like silicon, germanium, and gallium arsenide with and without ion implantation became the basis for the development of the devices that constitute the new industrial revolution. Developments in the CPA, outlined in chapter 5, have made it possible to understand measurements on alloys.

9.1 Incommensurate concentration waves

With the background outlined above, it is clear that experimental and theoretical researchers were impressed when a theory was put forward that explains a very subtle result of diffraction experiments. Diffraction patterns elucidate the phases that alloys transform into when the concentrations of the constituent elements change. Electron microscopy shows that alloys with concentrations that are off stoichiometry have domains that have the stoichiometric ordering and others that are simple solid solutions. Examples of ordered structure are Cu_3Au, CuAu, and $CuAu_3$. The palladium–rhodium system has a canonical clustering phase diagram. The first step in an alloy design project is to study the relevant phase diagrams.

The relevance to the present study of the pioneering work of Krivoglaz, Katchaturyan, Landau, Lifshitz, and others on the theory of ordering and clustering in alloys based on concentration waves is explained in two detailed papers by Gyorffy *et al* [3] and Stocks *et al* [4]. The books in which these papers are published are long since out of print, but they can be found in ebook form from Springer. As pointed out in these papers, the advantage in using concentration waves is that the k-vectors of the relevant waves are directly measurable by x-ray diffraction experiments.

The phase transformations in alloys are typically of first order, but there are a few points in the phase diagrams that describe second order phase transformations. The Landau approach is explained in modern books on statistical mechanics [5]. Landau theory starts from the concept of an alloy in the highest possible symmetry state in which every site in the crystal is the same. The probability of finding an A atom on the site is c and the probability of finding a B atom is $(1 - c)$. This is like the ensemble averaged alloy state underpinning CPA theory. Symmetry is broken in a real crystal by a concentration fluctuation described by assuming a measurable effective concentration c_i at each site i. The free energy associated with a specific fluctuation about the symmetry state is written

$$\delta F = \frac{1}{2!}\sum_{i,j}\gamma_{ij}^{(2)}\delta c_i\delta c_j + \frac{1}{3!}\sum_{i,j,k}\gamma_{ijk}^{(3)}\delta c_i\delta c_j\delta c_k + \dots, \qquad (9.1)$$

where

$$\delta c_i = c_i - c, \tag{9.2}$$

are the order parameters. The phenomenological coefficients $\gamma_{ij\ldots}^{(l)}$ are not related to a microscopic theory of atomic interactions and are arbitrary at this stage. The Landau approach is to use the macroscopic symmetries of the system to put conditions on them. The experimental concentration fluctuation is then the one that minimizes δF.

The concentration wave approach is a version of Landau theory that focuses on the first coefficient $\gamma_{ij}^{(2)}$, which depends only on the distance between sites, $\mathbf{R}_j - \mathbf{R}_i$. The eigenstates of

$$\sum_j \gamma_{ij}^{(2)} \delta c_j^v = \lambda_v \delta c_i^v, \tag{9.3}$$

describe the possible fluctuations, and the one corresponding to the lowest eigenvalue is the desired solution. Writing $\delta c_i(\mathbf{k}) = a_i(\mathbf{k}) e^{i\mathbf{k}\cdot\mathbf{R}_i}$, and $\gamma^{(2)}(\mathbf{k}) = \frac{1}{N}\sum_i e^{i\mathbf{k}\cdot(\mathbf{R}_i - \mathbf{R}_j)}\gamma_{ij}^{(2)}$ the lowest eigenvalue that is of the assumed form is the solution $\mathbf{k}_v$ of

$$\nabla_{\mathbf{k}} \gamma^{(2)}(\mathbf{k})|_{\mathbf{k}_v} = 0. \tag{9.4}$$

Lifshitz studied the forms that $\gamma_{ij}^{(2)}$ could take due to the interplay of the space group and the point groups of a specified Bravais lattice. He then solved equation (9.4) for the k-vectors that minimize $\gamma^{(2)}(\mathbf{k})$. The concentration waves corresponding to the special k-vectors obtained in this way describe all the known ordered structures for the actual alloys that have the specified Bravais lattice. For example, the Cu_3Au structure, based on the fcc Bravais lattice, is predicted by the special points (1, 0, 0), (0, 1, 0), and (0, 0, 1). The CuZn structure, based on the bcc Bravais lattice, is predicted by the special point (1, 1, 1). There would be no reason for the theory as developed to predict a special k-point that is incommensurate with the underlying lattice, although Landau theory can be generalized to do that.

Given the widespread successes of Monte Carlo methods and molecular dynamics calculations in statistical mechanics studies of condensed matter, it is not surprising that the next step in the theory of order–disorder transformations was to introduce interatomic potentials. The subscripts on the potentials $v_{ij}^{\alpha\beta}$ reference the distance between sites $\mathbf{R}_j - \mathbf{R}_i$, and the superscripts give the atomic species on the sites. Among other things, these potentials can be used to obtain the Landau parameters $\gamma_{ij}^{(2)}$, and hence link with the concentration wave analysis described above. There are two caveats that should be made about this process. In the first place, the $v_{ij}^{\alpha\beta}$ are no less phenomenological than the $\gamma_{ij}^{(2)}$. In the second place, the required physical picture is that the atoms in the alloy have all of their electrons in tightly bound orbits so that they appear to be charge neutral objects that see each other only through the $v_{ij}^{\alpha\beta}$. This picture may have some validity for highly ionic compounds, but it flies in the face of everything we know about metallic bonding from band theory calculations. Universality could be invoked to override these physical arguments on the basis that one model is as good as another as long as it has the necessary critical points. Nonetheless, very bizarre assumptions would have to be made about $v_{ij}^{\alpha\beta}$ to obtain incommensurate concentration waves.

As more precise x-ray diffraction measurements were made on alloys, they began to show data that could only be interpreted as concentration waves with k-vectors that are not commensurate with the underlying Bravais lattice. Other incommensurate waves are known to exist in low-dimensional structures. The k-vectors for these spin-density and charge density waves were successfully related to the Fermi surfaces. Using simple models, it was demonstrated that singularities in the response functions arise when the k-space integrals are over flat and parallel sheets of Fermi surface. The special point is the normal vector that measures the distance between the sheets. Thus, the incommensurate k-vectors are related to dimensions in reciprocal rather than real space.

It is easy to find flat parallel sheets of Fermi surface in one-dimensional structures, since that is the only shape that the Fermi surface takes. For that reason, spin and charge density waves are only studied in low-dimensional systems.

It is hard to imagine Fermi surfaces with flat parallel faces in alloys. The structures of disordered alloys are not invariant under the operations of a space group, so Bloch's theorem does not apply and k-vectors are not good quantum numbers. Also, there is no geometrical requirement that would lead to flat parallel faces as in low-dimensional structures. Studies of the CPA spectral density function $A(E, \mathbf{k})$ described in chapter 5 show that, at least in some cases, there will be sharp peaks that describe a slightly smeared out Fermi surface. It would then follow that alloys would be a fertile area in which to search for flat parallel sheets in k-space. If the sheets were convex at one concentration and concave at another, then there should be a range of concentrations inbetween in which they were flat and parallel.

Gyorffy and Stocks [6] investigated this possibility in copper–palladium alloys for which incommensurate waves had been observed. Figures 9.1–9.3 show the results of their calculations. They compare their calculations with the positions of the diffuse scattering spots in the electron diffraction data of Ohsima and Watanabe [7]. These spots are clear indications of incommensurate concentration waves.

The excellent agreement between the calculations of constant energy surfaces with the spectral density $A(E, \mathbf{k})$ obtained from the KKR-CPA theory of alloys and the results of diffraction experiments encouraged Gyorffy and Stocks to develop the theory further.

9.2 Correlations and order in alloy concentrations

From one point of view, the theory of electronic structure described in previous chapters leads to a rather boring picture of a solid. The model treated in band theory is a perfect crystal that transforms under the operations of a space group without the slightest error. There are no misplaced atoms or even thermal vibrations. The CPA theory of disordered alloys leads to a model that is perfect in a mathematical sense. There is an underlying Bravais lattice, just like in band theory. Statistically, every

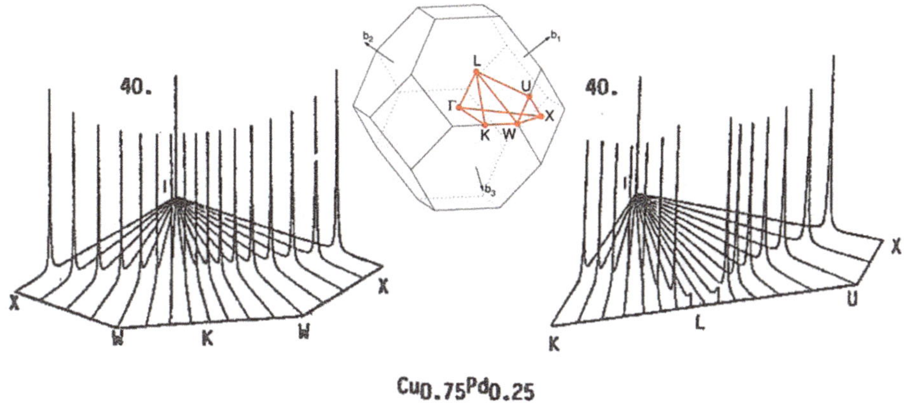

Figure 9.1. A sample of the calculations of the spectral density function for $Cu_{0.75}Pd_{0.25}$. The energy is set at the Fermi energy, E_F, and $A(E_F, \mathbf{k})$ is calculated for k-vectors along a line from the center, Γ, to the surface of the Brillouin zone. Calculations were done for many directions, but only a few on symmetry surfaces are shown. For this alloy, there are sharp peaks in the spectral zone that define a constant energy surface which comes into contact with the surface of the Brillouin zone at the point L, qualitatively similar to the Fermi surface of pure Cu. The intersection of the constant energy surface with the XWK plane illustrates that it is flat and defines a surface that is perpendicular to the ΓK (110) line. The critical k-point is two times the distance from Γ to the intersection of the plane with the ΓK line, $2k_F(0.25)$.

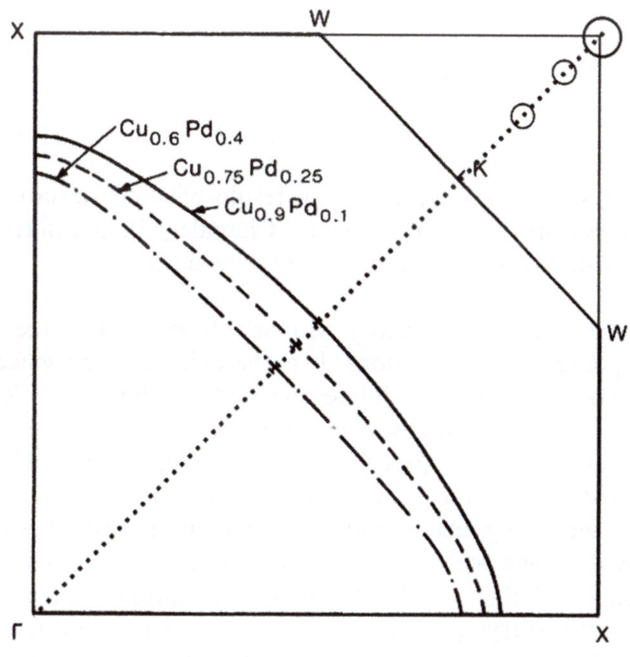

Figure 9.2. The intersections of constant energy planes for three different concentrations of copper–palladium. The distance from Γ to the end of the dotted line that is encircled is $\sqrt{2}$. That is also precisely twice the distance from Γ to the intersection with the $Cu_{0.9}Pd_{0.1}$ plane, $2k_F(0.1)$, and is a Lifschitz special point for the $L1_2$ structure. The points at the centers of the smaller circles are $2k_F(0.25)$ and $2k_F(0.4)$.

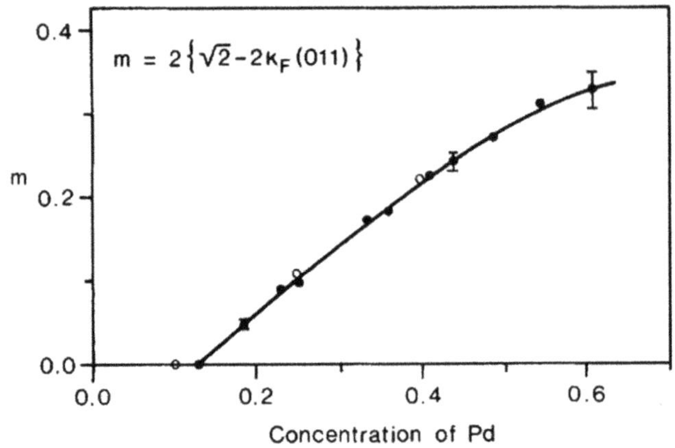

Figure 9.3. For technical reasons the experimental diffraction data is reported as the separation of the diffuse scattering spots $m = 2(\sqrt{2} - 2k_F(x))$, which is twice the distance from the end of the dotted line to the points indicated by the smaller circles in the preceding figure. The calculated points are shown by the open dots at concentrations of 0.1, 0.25, and 0.4. The experimental data are shown by the black dots. Reprinted figure with permission from [6]. © 1983 American Physical Society.

site is the same in that the probability of finding the A atom on a site is c, and the probability of finding a B atom is 1-c. This is the symmetric starting point for the Landau theory described in the previous section. It is the final result in the CPA, and it makes it possible to define a spectral density $A(E, \mathbf{k})$, which is periodic in k-space and has Brillouin zones.

Similar remarks could be made about the results of standard statistical mechanics applied to any system. As pointed out by Chandler [8], the purpose of modern developments in statistical mechanics is to go beyond this picture.

> 'Statistical mechanics is the theory with which we analyze the behavior of natural or spontaneous fluctuations. It is the ubiquitous presence of fluctuations that makes observations interesting and worthwhile. Indeed, without such random processes, liquids would not boil, the sky would not scatter light, indeed every dynamic process in life would cease. It is also true that it is the very nature of these fluctuations that continuously drives all things toward ever increasing chaos and the eventual demise of any structure. (Fortunately, the time scales for these eventualities are often very long, and the destruction of the world around us by natural fluctuations is not something worth worrying about.) Statistical mechanics and its macroscopic counterpart, thermodynamics, form the mathematical theory with which we can understand the magnitudes and time scales of these fluctuations, and the concomitant stability or instability of structures that spontaneous fluctuations inevitably destroy.'

In reference [3], Gyorffy *et al* derive a theory that allows for fluctuations in the concentrations of A and B atoms on the sites of a Bravais lattice. They start by justifying a free energy for a model that allows for the atom on a site to change from A to B or vice versa

$$\Phi(T,\{c_i\}) = E_e(\{c_i\}) + k_B T \sum_i [c_i \ln c_i + (1-c_i)\ln(1-c_i)] - \sum_i v_i(c_i - 1/2). \tag{9.5}$$

In this equation, $E_e(\{c_i\})$ is the total energy of all the electrons in a system of atoms whose arrangement is according to the set of one-site concentrations $\{c_i\}$. The second term is a simplified expression for the entropy of a system with the specified set of concentrations. The parameters v_i in the third term are a type of chemical potential that give the average energy cost of switching the atom on site i from A to B, or vice versa.

The set of concentrations, $\{c_i\}$ are the same ones used in the preceding section. It should be noted that they do not describe a specific arrangement of atoms, but only a tendency for an arrangement. It should also be pointed out that the parameters in the free energy are hypothetical. We know what they do, but we don't know how to calculate them. As the derivation evolves, approximations will be made that make it possible to specify them more precisely.

The minimization of the free energy leads to a set of conditions

$$\frac{\partial \Phi}{\partial c_i} = \frac{\partial E_e(\{c_i\})}{\partial c_i} + k_B T[\ln c_i - \ln(1-c_i)] - v_i = 0, \tag{9.6}$$

or

$$\begin{aligned} &S_i^{(1)}(\{c_i\}) + k_B T \ln \frac{c_i}{(1-c_i)} - v_i = 0 \\ &\frac{\partial}{\partial x} \ln \frac{x}{(1-x)} = \frac{1}{x} - \frac{\partial}{\partial x} \ln(1-x) \\ &= \frac{1}{x} - \frac{\partial}{\partial x} \ln(1-x) = \frac{1}{x} + \frac{1}{1-x} = \frac{1}{x(1-x)} \end{aligned} \tag{9.7}$$

with

$$S_i^{(1)}(\{c_i\}) = \frac{\partial E_e(\{c_i\})}{\partial c_i}. \tag{9.8}$$

The derivative of equation (9.7) then leads to

$$\frac{\partial S_i^{(1)}(\{c_i\})}{\partial c_j} + k_B T \frac{1}{c_i(1-c_i)} \delta_{ij} - \frac{\partial v_i}{\partial c_j} = 0, \tag{9.9}$$

or

$$S_{ij}^{(2)}(\{c_i\}) + k_B T \frac{1}{c_i(1-c_i)} \delta_{ij} - k_B T \frac{1}{q_{ij}^{-1}} = 0 \tag{9.10}$$

where $S_{ij}^{(2)}(\{c_i\})$ is called the direct correlation function

$$S_{ij}^{(2)}(\{c_i\}) = \frac{\partial S_i^{(1)}(\{c_i\})}{\partial c_j} = \frac{\partial^2 E_e(\{c_i\})}{\partial c_i \partial c_j}, \tag{9.11}$$

and

$$\frac{\partial v_i}{\partial c_j} = k_B T \frac{1}{q_{ij}^{-1}}. \tag{9.12}$$

The most profound of the equations above is the last one, equation (9.12). The meaning of the equation is that there is a matrix **q** whose elements, q_{ij}, are correlation functions with the meaning that they give the probability for finding an A atom on site j if we know that there is an A atom on site i. The quantity q_{ij}^{-1} is the ij element of the inverse of **q**. It is not at all obvious that the derivative in equation (9.12) should be proportional to that quantity. We will find that, when the equations have fully evolved, the v_i do not appear. A large amount of research in statistical mechanics has been dedicated to the study of correlation functions in liquids and gases, but very little on the arrangement of A and B atoms on a lattice. Gyorffy *et al* credit Evans [9] and also Krivoglaz [10] for leading them to this conclusion. Accepting this argument, equation (9.10) can be rewritten

$$q_{ij} = c_i(1 - c_i)\delta_{ij} - \beta c_i(1 - c_i)\sum_l S_{il}^{(2)}(\{c_i\})q_{lj}. \tag{9.13}$$

The next step in the argument is to assume that the correlations q_{ij} and $S_{ij}^{(2)}(\{c_i\})$ are proportional to the distance between the i and j lattice sites, $\mathbf{R}_j - \mathbf{R}_i$. This makes it possible to perform a lattice Fourier transform

$$q(\mathbf{k}) = c(1 - c) - \beta c(1 - c)S^{(2)}(\mathbf{k})q(\mathbf{k}), \tag{9.14}$$

or

$$q(\mathbf{k}) = \frac{c(1 - c)}{1 + \beta c(1 - c)S^{(2)}(\mathbf{k})}. \tag{9.15}$$

This equation is the goal of the derivation. In diffraction experiments, the diffuse scattering intensity between the Bragg peaks is proportional to the Fourier transform of the correlation function, $q(\mathbf{k})$. This is true for the gases and liquids of the alloys studied here. An analogous equation was derived earlier for alloys, but a classical interatomic potential model was assumed. Equation (9.15) is more general in that the origin of the energy is not restricted.

From LDA theory, it is known that the total electronic energy $E_e(\{c_i\})$ is the sum of the one-electron eigenvalues plus double counting terms

$$E_e(\{c_i\}) = \int_{E_B}^{E_F} E\rho(E)dE + \text{double counting}. \tag{9.16}$$

Using the fact that the DOS is simply the energy derivative of the IDOS, it is easy to show that the sum of eigenvalues is

$$E_{sp}(\{c_i\}) = E_F N(E_F) - \int_{E_B}^{E_F} N(E)dE. \tag{9.17}$$

The success in calculating the incommensurate concentration waves in copper–palladium, described in the preceding section, encouraged Gyorffy *et al* to turn to the CPA to calculate the electronic energy for the alloy.

In order to replace the general expressions above with their CPA counterparts, it is necessary to introduce the concept of an inhomogeneous CPA. The IDOS for the canonical CPA was given by the Lloyd formula in equation (5.15). In order to include fluctuations, the Lloyd formula is generalized for a hypothetical inhomogeneous CPA

$$\begin{aligned} N_I(E, \{c_i\}) = N_0(E) &- \frac{1}{\pi}\text{Im ln det}[\mathbf{M}_c - \mathbf{G}(E)] \\ &- \frac{1}{\pi}\sum_i \text{Im ln det}[c_i\mathbf{D}_{A,i} + (1 - c_i)\mathbf{D}_{B,i}], \end{aligned} \tag{9.18}$$

with

$$\mathbf{D}_{X,i} = [\mathbf{I} + (\mathbf{m}_{X,i} - \mathbf{m}_{c,i})\boldsymbol{\tau}_c^{ii}]^{-1}. \tag{9.19}$$

The matrix $\mathbf{M}_c$ has elements

$$M_{c,ij} = \mathbf{m}_{c,i}\delta_{ij}, \tag{9.20}$$

and $\mathbf{G}(E)$ has elements

$$G_{ij} = \mathbf{g}(\mathbf{R}_i - \mathbf{R}_j). \tag{9.21}$$

It should be pointed out that this formalism is totally different from the polymorphous CPA described in chapter 5. It is simply a mathematical artifice that makes it possible to use the CPA to calculate the direct correlation function

$$S_{ij}^{(2)}(\{c_i\}) = \frac{\partial^2 E_{sp}^{CPA}(\{c_i\})}{\partial c_i \partial c_j} = -\int_{E_B}^{E_F} \frac{\partial^2 N_I(\{c_i\})}{\partial c_i \partial c_j} dE. \tag{9.22}$$

After the derivatives have been taken, the IDOS is allowed to revert to the Lloyd formula in equation (5). The resulting equations are described in detail in Gyorffy *et al* [3].

Figure 9.4 shows the result of a calculation of the total scattering intensity for $Cu_{.75}Pd_{.25}$ using the formalism sketched above.

The experimental measurements of Ohsima and Watanabe are shown in figure 9.5.

The two preceding figures demonstrate the success of the electronic explanation for the structure of solids. The theory has gone beyond the association of the wavelength of the concentration waves with geometrical features of the Fermi surface to the prediction of the entire k-dependence of the diffraction intensity. The same approach, when applied to palladium–rhodium alloys gives the following prediction for the diffraction pattern.

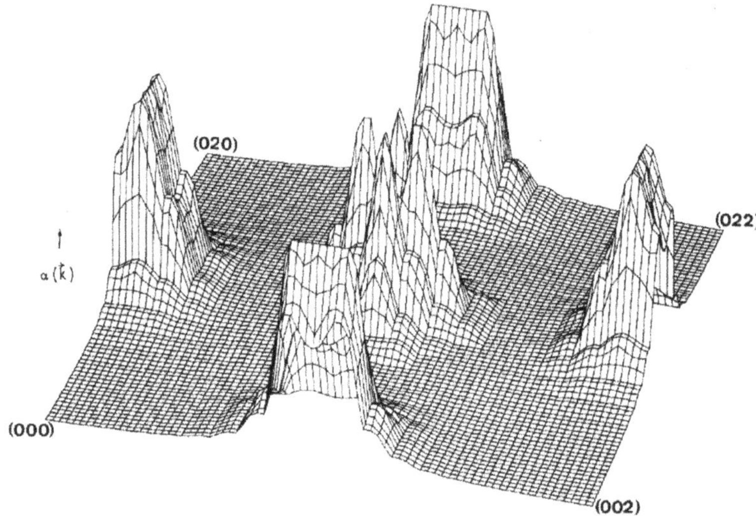

Figure 9.4. The calculated $q(\mathbf{k})$ for $Cu_{.75}Pd_{.25}$ using the equations in this section. The LDA double counting terms are left out, and the CPA is used. The large peaks at (001), etc are the Bragg peaks for the underlying fcc lattice. The cluster of four peaks around the (011) point are the result of the incommensurate concentration waves.

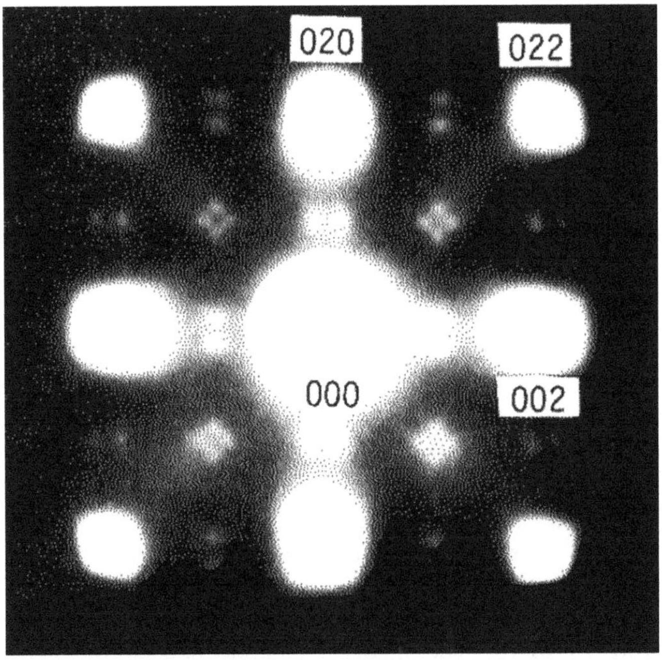

Figure 9.5. The electron diffraction data of Ohsima and Watanabe for $Cu_{.75}Pd_{.25}$. The calculations in the previous figure show only the upper right quadrant of this image. The large blobs are the Bragg peaks, and the smaller dots are the diffuse scattering peaks from the incommensurate concentration waves. Reproduced from [7] with permission of the International Union of Crystallography.

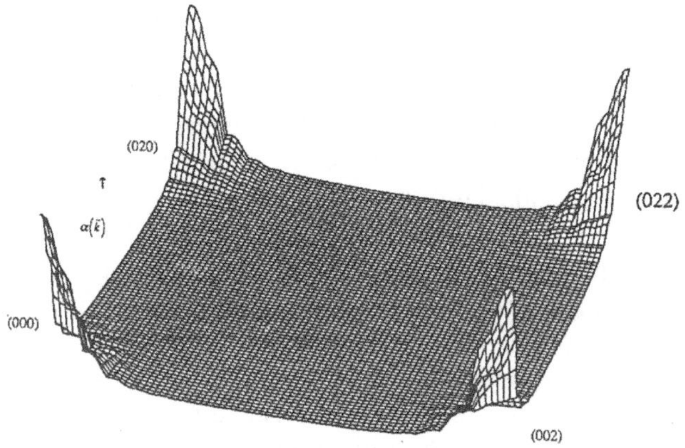

Figure 9.6. The calculated $q(\mathbf{k})$ for $Pd_{.25}Rh_{.75}$ using the equations in this section. The LDA double counting terms are left out, and the CPA is used. The large peaks at (000), (002), etc are the Bragg peaks for the fcc lattice. There are no superstructure peaks because this is a clustering rather than an ordering system.

The simple diffraction pattern shown in the preceding figure is observed in palladium–rhodium. The phase diagram shows a miscibility gap. The system will be discussed in greater detail in the next section, but the abbreviated story is that the diffraction pattern in figure 9.6 is in complete agreement with the experiment.

Gyorffy *et al* [3] delineated the basic ideas for the mean field concentration functional theory (MFCFT) of structural transformations in alloys outlined in this section. The basic idea is to replace semi-classical models that make use of unphysical atom–atom interactions with a fully electronic picture. The MFCFT has been and continues to be developed further. In particular, the importance of the double counting terms and the inclusion of a normalization factor called the Onsager cavity field are described in great detail in two papers by Staunton *et al* [11].

9.3 The embedded cluster Monte Carlo method

The crux of the Monte Carlo method for calculating the equilibrium distribution of atoms in an alloy is to obtain the energy required to replace an A atom on a particular site with a B atom when the configuration of the atoms on neighboring sites, κ, is specified

$$\delta H_\kappa = E_B(\kappa) - E_A(\kappa). \tag{9.23}$$

A random number between 0 and 1 is generated, and the atoms are interchanged if

$$z < e^{-\delta H_k/2k_BT}. \tag{9.24}$$

Conventionally, the interchange energies are obtained from an Ising-like Hamiltonian using interatomic potentials. We have argued against the use of such potentials, so we propose another method that calculates the δH_κ directly from the electronic structure of the alloy. Such a method exists, and is called the embedded cluster method (ECM) [12].

In the ECM, the electronic energy is calculated for a specified distribution of A and B atoms on the sites of the first few nearest-neighbor shells surrounding the central site. The effective scattering matrix $t_c(E)$ from the ordinary CPA is placed on all of the other sites. The difference in the total energy of two such systems differ only in that an A atom replaced with a B atom at the center of the cluster gives δH_κ.

KKR band theory calculations using the scalar relativistic approximation described in chapter 8 were used to find the theoretical lattice constants for pure palladium and rhodium by minimizing the calculated energies. The lattice constants for disordered palladium–rhodium alloys were calculated using the SRA version of the KKR-CPA. All of these lattice constants agree extremely well with the experimental values. Non-relativistic calculations do not agree with the experiment, which demonstrates the importance of relativistic effects for this system.

For temperatures above 1220 K, palladium and rhodium alloys are distributed randomly on the sites of a face-centered cubic (fcc) Bravais lattice. We construct our clusters based on that lattice, which has 12 nearest neighbors surrounding the central site. There are thus $2^{12} = 4096$ configurations κ of atoms in our clusters. Because of the cubic symmetry, only 144 clusters will give different values for δH_κ.

The 144 different interchange energies for the 50% alloy are shown in figure 9.7. They are grouped together by the number of rhodium atoms in the nearest-neighbor shell because in Ising-type calculations the exchange energy is linearly dependent on that variable. We calculated these energies within the frozen potential approximation, which means that only the sum of one-electron eigenvalues was used. Tests showed that this is an excellent approximation for the small energy differences involved.

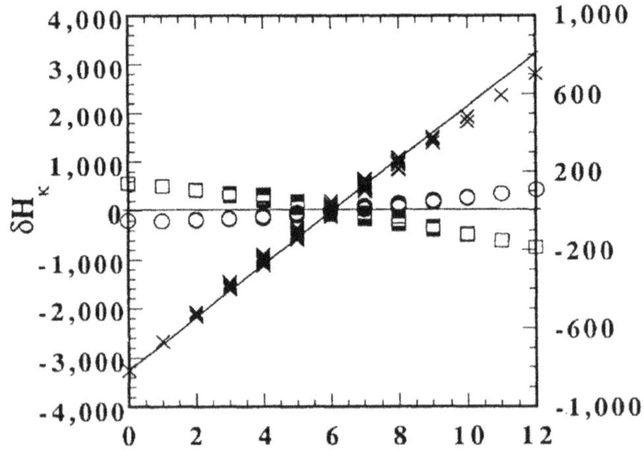

Figure 9.7. The x's are the interchange energies for a 50% palladium–rhodium alloy plotted as a function of the number of rhodium atoms in the nearest-neighbor shell. The open circles are the interchange energies for the 25% alloy minus those for the 50% alloy, the values being read from the right-hand scale. The open squares are the interchange energies for the 75% alloy minus those for the 50% alloy, the values being read from the right-hand scale. The energies are in degrees Kelvin. Reprinted figure with permission from [12]. © 1993 American Physical Society.

The amount that the δH_κ for the 25% and 75% alloys differs from that for the 50% alloy is quite small (note the change in scale in the figure), but it affects the phase boundary in an important way. The δH_κ were calculated for 10%, 25%, 50%, 75%, and 90% rhodium, and they can be found for any concentration of interest by interpolation.

By comparison with the straight line drawn in the figure as a guide to the eye, it can be seen that the δH_κ are not symmetrical about $n_{\mathrm{Rh}} = 6$ as they would be if they came from interatomic potentials. The other obvious difference is that there are several values of δH_κ for a given n_{Rh} because they depend on the precise configuration κ.

These δH_κ were used in a grand canonical Monte Carlo calculation on a sample made up of 55 296 atoms to obtain the phase boundary that is compared with the experiment in figure 9.8. The Monte Carlo calculation is of a standard form, although the asymmetry of the interchange energies introduces some difficulties that are not considered in the literature.

The slope of the δH_κ versus n_{Rh} curve has the general form that in a magnetic analogy would correspond to a ferromagnetic interaction. In an alloy, this means that the constituent atoms prefer to be surrounded with atoms of their own kind, which is called clustering. For this reason, the shape of the boundary shown in the figure, called a miscibility gap, was expected. Above the boundary, the system exists as a homogeneous substitutional alloy, while within the boundary, it exists as a mixture of two phases. One phase is a palladium-rich substitutional alloy and the other is rhodium rich, the concentrations for a given temperature being the boundary points on the miscibility gap. The amounts of the two phases depend on the concentration of the alloy, and are given by the lever rule. The highest point on the calculated miscibility

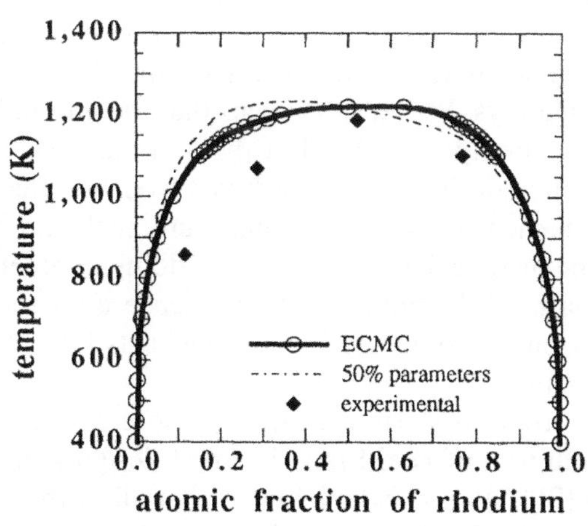

Figure 9.8. The phase boundary for the palladium–rhodium alloy system calculated with the ECMC is shown by the solid line, which is a cubic spline fit to the Monte Carlo results indicated by open circles. The temperatures at which four samples with the indicated concentrations were observed to undergo transitions to the two-phase region by Shield and Williams [13] are shown by the diamond-shaped points. The light dash-dotted line is the phase boundary calculated with interchange energies for the 50% alloy. Reprinted figure with permission from [12]. © 1993 American Physical Society.

gap has a temperature of 1220 K, which can be compared with the highest transformation temperature observed experimentally, 1190 K.

If the concentration dependence of the interchange energies is ignored and the phase boundary is calculated with the interchange energies for the 50% alloy, the asymmetry in these parameters is such that the highest point on the miscibility gap is on the palladium-rich side. When the concentration dependence of the δH_{κ} is included in the calculation, the peak of the miscibility gap is on the rhodium-rich side. This agrees with the experiment. The primary reason for the ultimate shape of the curve is the change in the lattice parameters with concentration.

In order to clarify the statements about the two-phase structure of the alloys below the miscibility gap, we show a cartoon version of the grains and grain boundaries that would be observed by an electron microscope.

The existence of grains and grain boundaries in a polycrystalline sample would obviously reduce the strength of the material. It would also increase the resistivity.

9.4 High entropy alloys

In the 1960s, followers of the original Star Trek series would frequently see Mr Spock try unsuccessfully to dent the surface of an alien ship and report something like 'Sensors show it is solid, but its composition is unknown to us.' Fans with scientific training would laugh at this because we knew that the elements in the periodic table are the same on every planet in the Universe, and we felt that everything we needed to know about combining elements to create structural materials had been worked out years earlier. The 2004 paper by Yeh *et al* [14] challenged our smug assumptions. This paper introduced the concept of high entropy alloys (HEAs). These are not simply new materials. The paper describes a new strategy for alloy design, and illustrates it with a collection of alloys that display measured hardness values that are two to three times larger than those for other widely used structural alloys. Perhaps the aliens that the intrepid Star Trek explorers had come across were simply way ahead of them in the development of HEAs.

The program for alloy design that has evolved over hundreds of years is to start with some principle element such as iron, aluminum, nickel, or (if you have enough money) titanium. One then begins the process of adding in other elements to achieve particular design goals. With centuries of experience and, today, new theoretical understanding based on quantum mechanics, scientists have become very good at carrying out this program.

The HEA design program starts with five or more principle elements in nearly equimolar ratios, such as CoCrFeMnNi. The first design goal is to create a single phase solid solution (SPSS) alloy, which means that all of the atoms are distributed randomly on the sites of a simple Bravais fcc or bcc lattice. An electron microscope would see no grains or grain boundaries like the ones in figure 9.9, a sketch of the grains and grain boundaries that would be observed by an electron microscope. For a 50% palladium–rhodium alloy at room temperature, roughly half of the grains would be almost pure palladium and the other half would be almost pure rhodium. The SPSS structure should immediately lead to alloys with desirable strength,

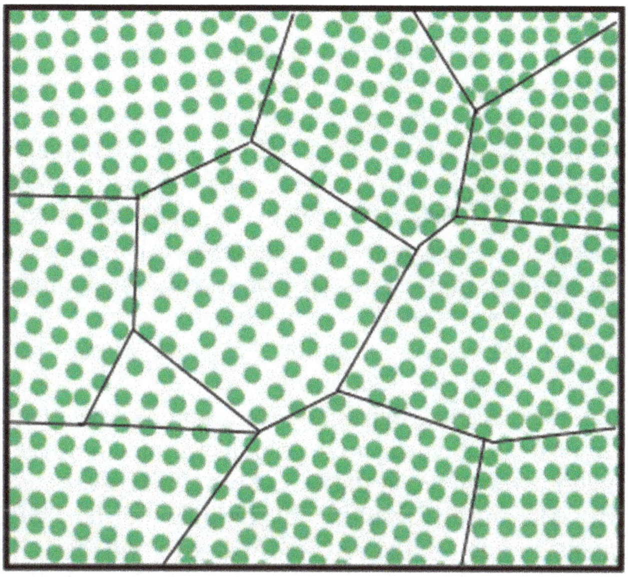

Figure 9.9. A sketch of the grains and grain boundaries that would be observed by an electron microscope. For a 50% palladium–rhodium alloy at room temperature, roughly half of the grains would be almost pure palladium and the other half would be almost pure rhodium.

corrosion resistance, imperviousness to annealing, low resistivity, etc. These properties can be further enhanced by the addition of other elements. The basic idea, and the origin of the name HEA, is that the high entropy of mixing would overwhelm the energetic effects (enthalpy) that would cause the alloy to develop ordered phases or cluster, as described in previous sections.

The theoretical model treated by the CPA of chapter 5 is a SPSS, but we showed that it can be used as a starting point for treating multiphase alloys. For example, the only way to get a SPSS in the palladium–rhodium system described in the previous section is to heat the material above 1220 K. At room temperature, as explained above, the alloy is an amalgamation of palladium-rich and rhodium-rich crystallites. One might say that the palladium–rhodium binary alloy is the opposite of a HEA.

It has been shown [15] that the hypothesis that the underlying principle of HEAs is the high entropy of mixing is limited. The experiment was to start with the SPSS CoCrFeMnNi and remove and replace elements one at a time. The entropy remains the same in all samples, but some remain SPSS while others transform into multiphase materials. The authors conclude that high configurational entropy provides a way to rationalize a HEA, but it is not a useful *a priori* predictor of which alloys with multiple principle elements will form thermodynamically stable SPSS.

One of the technical problems facing the development of these alloys is the sheer number of possible combinations of elements taken five, six, or seven at a time that might lead to useful HEAs. For this reason, theoretical methods based on quantum mechanics are imperative.

Enthalpy calculations for binary systems exist as tables or databases that can be searched on the web. The calculations are simple enough in principle. The lattice constants and binding energy relative to separated atoms are calculated for single elements. Then the relaxed structures and binding energies for compounds that can be made from pairs of elements are calculated. By interpolating between these data, predicted enthalpies can be found for a wide range of binary systems.

Troparevsky *et al* [16] used the available enthalpy calculations for binaries supplemented by quantum mechanical calculations of their own to develop criteria for predicting the combinations of five, six, or seven elements that will form SPSS alloys. The basic idea can be found from figure 9.10.

The extremum enthalpies of formation δE_{AB} are found from the quantum mechanical calculations for the 36 couplets of elements that are formed from the 9 elements used in the preceding figure. If δE_{AB} for a given couplet is more negative than a prescribed limit, the elements in that couplet should not be included together in the test sample. For example, in the CoCrFeMnNi HEA, Fe could be removed and replaced with Mo and the resulting alloy would still be a SPSS. However, if some other element is removed and replaced with Mo, the δE_{FeMo} would cause the sample to be multiphase. Also, the immiscibility of Cu with the other elements indicated by the positive values for δE_{CuB} means that samples containing that element will not be SPSS. Using the data from figure 9.10—a plot of the extremum enthalpies of formation of the 36 binary alloys that could form from the 9 elements considered in reference [15]—all of the experimental observations in reference [15] can be explained.

In the web data, supplemental to this paper [17], results like the ones in the preceding figure are tabulated for the 435 couplets that can be formed from 30

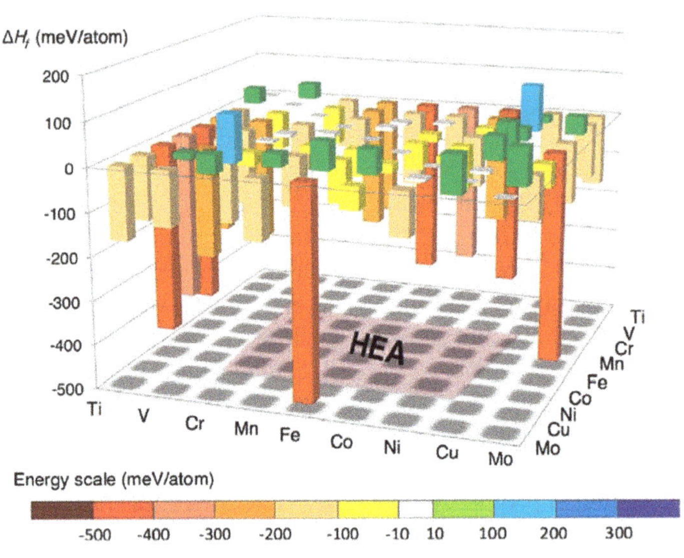

Figure 9.10. Plot of the extremum enthalpies of formation of the 36 binary alloys that could form from the 9 elements considered in reference [15]. Reproduced from [16] CC BY 3.0.

elements. These data make it possible to consider 142 506 five element samples, 593 775 six element samples, or 2 035 800 seven element samples. Knowledge of the δE_{AB} makes it possible to eliminate many of those, but there remain many possible HEAs to be found. Some of these may well have more desirable properties than the ones found so far.

The data on 30 elements represents a lower bound on all possibilities. Also, other minor elements can be added to fine tune a HEA for a particular purpose. It is clear from the number of possible HEAs that it will be necessary to find more criteria to focus on the most promising candidates.

References

[1] Smith M C 1956 *Principles of Physical Metallurgy* (Harper)
[2] Harrison W A and Webb M B 1960, *The Fermi surface: Proc. Int. Conf.* (*August 22–24, 1960, Cooperstown, New York*) (New York: Wiley)
[3] Gyorffy B L *et al* 1989 *Alloy Phase Stability* ed G M Stocks and A Gonis (Kluwer Academic Publishers) p 421
[4] Stocks G M *et al* 1994 *Statics and Dynamics of Alloy Phase Transitions* ed P E A Turchi and A Gonis (New York: Plenum Press) p 305
[5] Sethna J P 2006 *Statistical Mechanics: Entropy, Order Parameters and Complexity* (New York: Oxford University Press)
[6] Gyorffy B L and Stocks G M 1983 *Phys. Rev. Lett.* **50** 374
[7] Ohsima K and Watanabe D 1972 *Acta Crystallogr. Sec.* A **29** 520
[8] Chandler D 1987 *Introduction to Modern Statistical Mechanics* (New York: Oxford University Press)
[9] Evans R 1979 *Adv. Phys.* **28** 143
[10] Krivoglaz M A 1969 *Theory of x-ray and Thermal Neutron Scattering by Real Crystals* (New York: Plenum)
[11] Staunton J D, Johnson D D and Pinski F J 1994 *Phys. Rev.* B **50** 1450
Staunton J D, Johnson D D and Pinski F J 1994 *Phys. Rev.* B **50** 1473
[12] Wang Y, Faulkner J S and Stocks G M 1993 *Phys. Rev. Lett.* **70** 3287
[13] Shield J E and Williams R K 1987 *Scr. Met.* **21** 1475
[14] Yeh J W *et al* 2004 *Adv. Engr. Mater.* **6** 299
[15] Otto F *et al* 2013 *Acta Mater.* **61** 2628
[16] Troparevsky M C *et al* 2015 *Phys. Rev.* X **5** 011041
[17] http://link.aps.org/supplemental/10.1103/PhysRevX.5.011041

Chapter 10

Conclusions: beautiful minds

Writing a book is not a lot of fun, but it has the pleasurable feature that it provides an opportunity to reflect on the achievements of some of the most beautiful minds in your field. Three such minds, Jan Korringa, Walter Kohn, and Balazs Gyorffy, contributed enormously to the origination and development of multiple scattering theory, the local density approximation, and the application of these tools to achieve scientific explanations for previously unexplained experimental observations. They are no longer with us, but their contributions remain.

Some might imagine that, since all of this work is based on the Schrödinger or Dirac equations, the intellectual challenges inherent in the kind of theory that is the subject of this book are not as great as those faced by researchers who attempt to go beneath the elementary equations. The reason that this so-called reductionist hypothesis is not correct is best explained by another beautiful mind in the field of physics, Phil Anderson [1]. He points out the apparent inevitability to go on uncritically to what appears at first sight to be an obvious corollary of the reductionist hypothesis, which is: if everything obeys the same fundamental laws, then the only scientists who are studying anything really fundamental are those who are working on those laws. The main fallacy in this kind of thinking is that the reductionist hypothesis does not by any means imply a 'constructionist' one. The ability to reduce everything to simple fundamental laws does not imply the ability to start from those laws and reconstruct the Universe.

Anderson points out in his famous article 'More is Different':

> 'The constructionist hypothesis breaks down when confronted with the twin difficulties of scale and complexity. The behavior of large and complex aggregates of elementary particles, it turns out, is not to be understood in terms of a simple extrapolation of the properties of a few particles. Instead, at each level of complexity entirely new properties appear, and the understanding of the new behaviors requires research which I think is as fundamental in its nature as any other.'

doi:10.1088/2053-2563/aae7d8ch10

This view can be expressed in a less didactic fashion. One can stare at the Dirac equation forever and never predict the phenomenon of superconductivity. In 1950, no expert in either science or technology was predicting the ubiquitous dependence of humans all over the world on devices made with silicon based chips. Physicists are finding that their skills can be put to very good use in the field of medicine, to the surprise of many. More to the point of the present book, magnetism in iron and its alloys has been known for thousands of years, but it cannot be explained without the highly detailed calculations described in chapter 5. One could speculate that the incommensurate concentration waves observed in certain alloys might have something to do with their electronic structure, but that speculation was only given substance by the studies reported in chapter 9. The high entropy alloys that are being hailed for their uniquely useful properties are a true twenty-first century material. The name was not known before 2002. The concepts in this book have been shown to be very useful for predicting alloy compositions that will fall in this category, as discussed in the preceding chapter.

There are always people with some claim to expertize who state that science is dead and there is nothing left to discover. This kind of statement flies in the face of the experience of condensed matter physicists. The next big thing is probably lurking around the next corner. It is hoped that the theory in this book can be helpful in understanding it.

Reference

[1] Anderson P W 1972 *Science, New Series* **177** 393

Lightning Source UK Ltd.
Milton Keynes UK
UKHW050726020720
365855UK00003B/35

9 780750 314886